T0329752

Innovation Networks: Theory and Practice

NEW HORIZONS IN THE ECONOMICS OF INNOVATION

Series Editor: Christopher Freeman, *Emeritus Professor of Science Policy, SPRU – Science and Technology Policy Research, University of Sussex, UK*

Technical innovation is vital to the competitive performance of firms and of nations and for the sustained growth of the world economy. The economics of innovation is an area that has expanded dramatically in recent years and this major series, edited by one of the most distinguished scholars in the field, contributes to the debate and advances in research in this most important area.

The main emphasis is on the development and application of new ideas. The series provides a forum for original research in technology, innovation systems and management, industrial organization, technological collaboration, knowledge and innovation, research and development, evolutionary theory and industrial strategy. International in its approach, the series includes some of the best theoretical and empirical work from both well-established researchers and the new generation of scholars.

Titles in the series include:

The Strategic Management of Innovation
A Sociological and Economic Theory
Jon Sundbo

Innovation and Small Enterprises in the Third World
Edited by Meine Pieter van Dijk and Henry Sandee

Innovation, Growth and Social Cohesion
The Danish Model
Bengt-Åke Lundvall

The Economics of Power, Knowledge and Time
Michèle Javary

Innovation in Multinational Corporations in the Information Age
The Experience of the European ICT Industry
Grazia D. Santangelo

Environmental Policy and Technological Innovation
Why Do Firms Adopt or Reject New Technologies?
Carlos Montalvo Corral

Government, Innovation and Technology Policy
An International Comparative Analysis
Sunil Mani

Innovation Networks
Theory and Practice
Edited by Andreas Pyka and Günter Küppers

Systems of Innovation and Development
Evidence from Brazil
Edited by José E. Cassiolato, Helena M.M. Lastres and Maria Lucia Maciel

Innovation Networks Theory and Practice

Edited by

Andreas Pyka[1] and Günter Küppers[2]

[1]*Assistant Professor, University of Augsburg, Germany*
[2]*Director, Institute for Science and Technology Studies, University of Bielefeld, Germany*

NEW HORIZONS IN THE ECONOMICS OF INNOVATION

Edward Elgar
Cheltenham, UK • Northampton, MA, USA

Published by
Edward Elgar Publishing Limited
Glensanda House
Montpellier Parade
Cheltenham
Glos GL50 1UA
UK

Edward Elgar Publishing, Inc.
136 West Street
Suite 202
Northampton
Massachusetts 01060
USA

This book has been printed on demand to keep the title in print.

A catalogue record for this book
is available from the British Library

Library of Congress Cataloguing in Publication Data
Innovation networks : theory and practice / edited by Andreas Pyka and Günter Küppers.
p. cm. — (New horizons in the economics of innovation)
"Presents the results of an international research project on 'Simulating self-organizing innovation networks' (SEIN) funded by the European Community under the TSER Programme, contract #SOEI-CT-98-1107"—P..
Includes index.
1. Technological innovations—Case studies. 2. Business networks—Case studies. 3. Knowledge management—Case studies. I. Pyka, Andreas. II. Küppers, Günter. III. Series.

HD45 .I5377 2003
658.4'063—dc21 200209820

ISBN 1 84376 040 1 (cased)

Contents

List of Figures

List of Tables

List of Contributors

Ahrweiler, Petra, Dr., Institute for Political Sciences, University of Hamburg, Sedanstrasse 19, D-20146 Hamburg, Germany,
e-mail: ahrweiler@sozialwiss.uni-hamburg.de

de Jong, Simone, MERIT, University of Maastricht, P.O. Box 616, NL-6200 MD Maastricht,
e-mail: Simone.deJong@MERIT.UNIMAAS.NL

Gilbert, Nigel, Professor Dr., Department of Sociology, University of Surrey, GB-Guildford GU2 5XH, United Kingdom,
e-mail: Nigel.Gilbert@soc.surrey.ac.uk

Küppers, Günter, Dr. Privat Dozent, Institut für Wissenschafts- und Technikforschung, Universität Bielefeld. Universitätsstrasse 25, D-33615 Bielefeld, Germany,
e-mail: guenter.kueppers@uni-bielefeld.de

Pyka, Andreas, Dr., Lehrstuhl für Innovationsökonomik, Institut für Volkswirtschaftslehre, WiSo-Fakultät, Universität Augsburg, Universitätsstrasse 16, D-86159 Augsburg, Germany,
e-mail: andreas.pyka@wiwi.uni-augsburg.de

Saviotti, Pier Paolo, Professor Dr., Université Pierre Mendès France, B.P. 47, F-38040 Grenoble cédex 9,
e-mail: saviotti@grenoble.inra.fr

Vaux, Janet, Dr., Sociology Department, University of Bristol, 12 Woodland Road, GB-Bristol BS8 1 UQ,
e-mail: j.vaux@britishlibrary.net

Weber K. Matthias, Dr., Austrian Research Centres, Systems Research Technology – Economy – Environment, A-2444 Seibersdorf, Austria,
e-mail: matthias.weber@arcs.ac.at

Windrum, Paul, Professor Dr., Department of Business Information Technology, The Business School, Manchester Metropolitan University, Aytoun Building, Aytoun Street, GB-Manchester M1 3 GH, UK,
e-mail: paul.windrum@man.ac.uk

Project Partners

Petra Ahrweiler, Institute for Science and Technology Studies (IWT), University of Bielefeld, Germany

Chris Birchenhall, Maastricht Economic Research Institute on Innovation & Technology, Maastricht, The Netherlands

Wilma Coenegrachts, Maastricht Economic Research Institute on Innovation and Technology, Maastricht, The Netherlands

Nigel Gilbert, Centre for Research on Simulation in the Social Sciences, University of Surrey, Guildford, UK

Matthias Groß, Institute for Science and Technology Studies (IWT), University of Bielefeld, Germany

Stéphane Isoard, Joint Research Centre of the European Commission, Institute for Prospective Technological Studies, Seville, Spain

Anke Jobmann, Institute for Science and Technology Studies (IWT), University of Bielefeld, Germany

Uli Kowol, Institute for Science and Technology Studies (IWT), University of Bielefeld, Germany

Sylvie Micheland, Institut National de la Recherche Agronomique, Unité d'Economie et Sociologie Rurales, Grenoble, France

Sabine Paul, Institute for Science and Technology Studies (IWT), University of Bielefeld, Germany

Andreas Pyka, Institut für Volkswirtschaftslehre, WiSo-Fakultät, Universität Augsburg, Germany

Glen Ropella, Swarm Corporation, Santa Fé, CA, USA

Paolo Saviotti, Institut National de la Recherche Agronomique, Unité d'Economie et Sociologie Rurales, Grenoble, France

Monika Suri, Maastricht Economic Research Institute on Innovation and Technology, Maastricht, The Netherlands

Janet Vaux, University of Surrey, Guildford, UK

K. Matthias Weber, Joint Research Centre of the European Commission, Institute for Prospective Technological Studies, Seville, Spain

Paul Windrum, Maastricht Economic Research Institute on Innovation and Technology, Maastricht, The Netherlands

Zhijia Zhou, Institute for Science and Technology Studies (IWT), University of Bielefeld, Germany

PROJECT CO-ORDINATOR

Günter Küppers, University of Bielefeld, Germany

Acknowledgement

This book presents the results of an international research project on 'Simulating Self-Organising Innovation Networks' (SEIN) funded by the European Community under the Targeted Socio-Economic Research (TSER) Programme, contract #SOEI-CT-98-1107. The project team gratefully acknowledges this supportive funding.

PART ONE

Theoretical Background

1. The Self-Organisation of Innovation Networks: Introductory Remarks

Günter Küppers and Andreas Pyka

The term 'knowledge society' has taken on an important role in public discourse in the last few years as an indicator of a fundamental change in society (Stehr 1994). Originally, the term was introduced by Fritz Machlup (1984) to mark the increasing importance of knowledge for economic progress and development. Knowledge is beginning to acquire the status of a resource like matter or energy. Bell (1973) goes even further and speaks about a revolutionary change from industrial to post-industrial society. His connotation offers a promise of modernisation with knowledge taking on the central role for societal progress. The production, distribution and use of knowledge characterises the new epoch and becomes the driving force for modern knowledge societies.

The literature on economic history has long used the term 'science-based industries' to point to the exceptional importance of (scientific) knowledge in developing new processes and products. Even in the second half of the nineteenth century, industrial progress, particularly in the chemical dye and electrical industries, was closely tied to the availability of scientific and technical knowledge, and this led to the emergence of industrial research. By the end of the Second World War at the latest, science had permeated all walks of life and has now become the major source of technological and societal innovation processes.

Nowadays, the competitiveness of modern companies depends decisively on access to the latest knowledge and its application in innovation processes. The provision and implementation of new knowledge has become the central challenge facing industry. This is acknowledged in the term 'knowledge economy'. 'The problem of innovation is to process and convert information from diverse sources into useful knowledge about designing, making and selling new products and processes' (Freeman 1991, p. 501).

This increasing dependency of modern innovation processes on knowledge is only one aspect of the contemporary economy. A second and related aspect of innovation dynamics is the fact that innovations are becoming increasingly more complex. The history of the development and manufacture of new products can be broken down roughly into three phases that are now juxtaposed, although they historically formed a temporal sequence. The first phase is the so-called craft production, in which highly skilled workers produce customised products for individual customers. The second is the development of mass manufacturing, its characteristic the division of a product into its component parts, which are standardised, thus allowing greater efficiency in production (Rycroft and Kash 1999, p. 12). Since the second half of the twentieth century a new, third phase has emerged characterised by the production of complex products; that is, products consisting of an increasing number of parts interrelated inseparably in a dynamic relationship. Compared with earlier motor vehicles for instance,

> today's cars are complex sets of interacting automatic subsystems. For example, the fuel system has incrementally become an intricate linkage of sensors, microprocessors, and controls that govern the fuel injection process with such sensitivity that many young drivers have never known a car to cough and jump when it was cold. . . . The Ford Taurus has more computing power than the original Apollo that went to the moon [in 1969] (Rycroft and Kash 1999, p. 4).

Unlike the complexity of structures, in which the integration of various elements is treated as a pattern,[1] the concept of a dynamic complexity relates to a functional whole of a large number of interrelated processes or interactive elements. It is important to note that it is not the number of processes or elements involved which is the source of complexity. On the contrary, it is the quality of the linkage by which both the processes or elements are interwoven, allowing a new quality, that is, a new function, to emerge. In this respect also the notion of architectural innovation (Henderson and Clark 1990) is used. Therefore, a machine is not complex because it consists of various parts, but, at most, because it integrates the dynamics of these parts and unifies them in one function. It is precisely in this sense that modern products have become complex, as the example of fuel injection has shown.

Therefore, innovations are technologically complex because

- new products integrate many elements into few functions,
- their production requires the integration of many resources within one organisation and
- their realisation needs the integration of many individual users into only a few contexts of applications.

In addition, modern technologies have to meet economic, political and social demands. In all, today's new products have to:

- *Be economically viable*. In other words, it has to be possible to develop and produce new products with the available commercial resources. They have to be able to assert themselves on the market and they also have to make a profit.
- *Be technologically feasible*. This means they have to function in technological terms while simultaneously being safe and reliable.
- *Fit into the political landscape*. In other words, they must not contravene existing environmental and safety standards.
- *Be accepted by customers*. This means they must fit into a real application context and meet customers with the respective competencies allowing their wide application and use.

These functions cannot be met independently of each other. Changes to comply with technological safety may pose a risk to economic viability, new environmental standards may endanger technological functionality and technological specifications may threaten customer acceptance. Again, these functions have to be integrated in one product. As a result more dimensions of complexity arise. Because of the functional unity of a new product with respect to these different dimensions of complexity, the whole cannot be broken down into its component problems and partial solutions cannot be integrated into a new product. In general, complexity requires an integrated, multidisciplinary approach.

> This complexity has meant that multidisciplinary knowledge has become necessary for the generation and development of new products. In the computer industry, for example, the disciplines involved in the innovation process may range from solid state physics to mathematics, and from language theory to management science (Malerba 1992).
> So-called *go-it-alone* strategies or *conservative* strategies which mean that a firm relies only on its own R&D endeavours, cannot be successful in such a complex environment. Because of the *systemic character* of present-day technological solutions, technological development necessarily becomes a complex interactive process involving many different ideas, and their specific interrelationships (Pyka and Saviotti 2000, p. 13).

As a new characteristic of this 'new way of knowledge production' transdisciplinarity appears (see Gibbons et al. 1994). Network structures link the diverse knowledge of producers, suppliers and users located in different organisations in order to facilitate rapid exchange and decision making. 'In this light networks represent a mechanism for innovation diffusion through collaboration and the interactive relationship becomes not only a co-ordination device to create resources, but an essential enabling factor of technical progress' (Zuscovitch and Justman 1995).

Thus, innovation networks emerge as a new form of organisation within knowledge production. Innovation networks are considered

> to have three major implications: First, they are seen as an important co-ordination device enabling and supporting *inter-firm learning* by accelerating and supporting the diffusion of new technological know-how. Second, within innovation networks the *exploitation of complementarities* becomes possible, which is a crucial prerequisite to master modern technological solutions characterised by complexity and a multitude of involved knowledge fields. Thirdly, innovation networks constitute an organisational setting which opens the possibility of the *exploration of synergies* by the amalgamation of different technological competencies. By this, innovation processes are fed with new extensive technological opportunities which otherwise would not exist, or whose existence would at least be delayed (Pyka and Saviotti 2000, p. 15).

Little is known about innovation networks. Open questions are: What characterises an innovation network in comparison with classical forms of organisations? What is the structure of an innovation network, what are its elements, what are their basic interactions, what co-ordination mechanisms are important and what dynamics emerge from internal interactions? What is the relationship between an innovation network and its environment and how does the environment influence its dynamics?

This book has set itself the task of contributing to the understanding of such innovation networks and of doing this on several levels: first, empirically existing innovation networks are examined as case studies in various innovation domains in order to focus on completely different characteristics in the task assignment, the institutions involved or the form of co-operation. The individual case studies focus on biotechnology, which is essentially about new products (product innovation), on Virtual Centres of Excellence (VCE), which analyse a new system technology (new products and supporting systems), on Combined Heat and Power technology (CHP), which is concerned with transforming an old service supply system (both product and service innovations embedded in an architectural innovation) and on the design of web pages, which concentrates on the co-evolution of technology and innovation networks in e-commerce (producer-user networks).

The second goal of this book is to analyse findings from the case studies in order to obtain a general model for innovation networks that may then serve as the starting point for a computer simulation. The concept of self-organisation is shown to be a fruitful framework for this theoretical analysis. Finally, the computer simulation should permit the quasi-experimental study of the impact of critical factors and environmental influences on the structure and dynamics of innovation networks. The simulation should help to contribute to a general theory of innovation networks.

THE THEORETICAL APPROACH

Innovations emerge because of the interaction between scientific, economic and political systems. Formerly, the modes of interaction of these different systems tended to be confined to the national level. The concept of a national innovation system (NSI) captures the interactions of different institutions and organisations that create and adopt innovations in a country. Recently, the interactivity of innovation systems seems to have increased both at the international level ('globalisation'), and at lower levels of aggregation, moving down to regional and local level. In the meantime, inter-institutional collaborative agreements (inter-firm, firm-university, etc.) – previously judged by economists as being inefficient or at least unstable in the long run – have increased enormously in frequency and duration, especially, although not exclusively, in high-tech sectors such as new materials and biotechnology.

This increasing interactivity within and among several aggregation levels is related to the growing knowledge intensity of industrialised economies. From a theoretical point of view, this has two consequences: first, although networks have always been involved in bringing out innovations, their growing importance cannot be neglected anymore. Second, we must improve our understanding of the dynamics of knowledge creation and utilisation. In terms of the latter issue, such dynamics seem to have undergone a major structural change. Gibbons et al. (1994) maintain that a new mode of knowledge generation – called Mode 2 – has been created in addition to the more traditional Mode 1, which is based on a sharp distinction between science and technology and on the disciplinary organisation of the creation of knowledge. By contrast, Mode 2 comprises multidisciplinary, problem-oriented research, which is carried out by actors with heterogeneous interests, funded by multiple sources (public/private), and produced in multiple forms (texts, artefacts, new forms of regulations, etc.).

Here, we define innovation networks as interaction processes between a set of heterogeneous actors producing innovations at any possible aggregation level (regional, national, supranational). As such, an innovation network is a self-maintaining social structure created in an unstable situation because the actors involved (firms, universities, government agencies) have not been able to define either the innovation problem or its solution. This unstable situation is caused by the perception of uncertainty about what the innovation is and how it can be produced. It is equivalent to the situation of non-equilibrium which is decisive for natural self-organisation: non-equilibrium presses for compensation, and this compensation changes the non-equilibrium. This mutual change of non-equilibrium and compensation leads to a steady state, a dynamic equilibrium in which non-equilibrium and compensation condition each other and a state of dynamic order is achieved (see Küppers 1996). In the phenomenal domain of the social, the mechanism of self-organisation is again circular causality, that is, the circular feedback of causes and their effects. Instead of non-equilibrium and compensation, social self-organisation is driven by the mutual relationship of the perception of uncertainty and social interactions which aim to reduce it. Again a steady state is reached if both the perception of and the engagement with uncertainty condition each other (see Küppers 1999).

This conceptualisation stresses that innovations result from an ongoing process of mutual adjustment of uncertainty perception about the technical feasibility, the social acceptance, and the economic success of an invention. The complexity of the innovation problem is not – as many authors claim it to be – a consequence of the many dimensions of the innovation problem or of the numerous parts integrated in an innovation. It is the interrelatedness of these dimensions or parts respectively which makes the production of an innovation a complex problem. More precisely: it is its non-linearity which becomes the source of complexity. Little changes in one of the variables may produce strong changes in the others. This 'may produce' instead of 'produce' is the source of non-determinism, even within the deterministic laws of nature. This problem of non-linearity is the characteristic feature of the innovation problem: for example, the slightest technical change may improve the economic success of an innovation dramatically, yet it may also reduce its social acceptance in a fundamental way.

Because of this complexity, innovations cannot be produced in a traditional way: first, all technical problems are solved; second, a cost-effective way of manufacturing is found; and third, the innovation is successfully introduced into the market. To the contrary, all these problems must be solved in strong (non-linear) relationship to each other. Because each of these problem areas needs specific expertise, a solution can only be found if all the experts collaborate in defining the overall problem and in finding a solution. This is the essential element of innovation networks.

Innovation networks represent forms of social self-organisation that are developed or deployed to deal with the complexity of an innovation. They are operationally closed in the sense that the perception of complexity within networks generates social activities aimed at reducing complexity by integrating the different dimensions of the innovation problem. To this end there are no predetermined plans of action, nor measures for assessing the amount of complexity at any given point in time. An innovation has been found once all those involved in a network believe that the outcome satisfies the implicit assumptions of technical functionality, social appropriateness and economic profitability. Complete certainty is, however, not possible. Only after the successful introduction of a new product or a new service onto the market is it really known whether such assumptions are justifiable.

Whilst innovation networks are, because of their circular causality, operationally closed social systems, they are nevertheless open to those resources required for their operations, namely, the various knowledge claims required for the solution of innovation problems. Knowledge resources can be made available to innovation networks in a variety of ways: as information from databases and information systems, or as the expertise of participating members. Such expertise guarantees access to the embedded forms of firm-specific (tacit) and technology-specific (local) knowledge that are equally important to the processes of innovation. Given that innovation networks have to integrate a variety of competencies, they cannot afford to restrict membership. Rather, anybody who has something to contribute can become a member of such networks. Because of this, innovation networks differ from traditional organisations in which membership is typically contractual and include both more formal relationships such as research joint ventures, etc., as well as informal networks relying on mutual co-operation and learning (Cantner and Pyka 1998). Innovation networks have been developed to overcome the organisational restrictions or limits in dealing with innovation problems.

If self-organisation theory provides the adapted description of innovation networks, empirical studies relying on this approach have to cope with the complexity of their target, namely various changing parameters, changing combinations, growing and declining relations. What method of sociological research could be fit for investigating this area, taking into consideration the suggested theoretical concepts? The setting of an artificial social experiment, which allows study of the dynamics of innovation processes cannot be carried out in reality. For that reason we will argue for computer simulation as a tool for investigation of such complex systems. By intense sensitivity analysis and 'playing around' with parameters the complexity may be reduced and answers with respect to the structure and the dynamics of innovation networks can be found.

The different case studies analysed within this book are used as an empirical basis for the development of such a simulation model. One important dimension of innovation networks is their internal complexity. One reason for this complexity is the heterogeneity or homogeneity of their members. Networks in which only scientists from different disciplines collaborate are much less complex than those where the collaboration includes people from outside – people from industry, from other groups of society and so on. This internal complexity affects in a tremendous way the internal dynamics of the network.

A second dimension which is important for the dynamics of an innovation network is its foundation. It makes a great difference whether the network is installed from outside. Even if the network is going to operate autonomously its dynamics can be distinguished from those which emerge because some people decide within a complex problem to co-operate. The case studies chosen for our analysis can be ranked by their increasing complexity; they cover a broad landscape of innovation networks which are of empirical relevance.

THE CASE STUDIES

Four case studies have been chosen to give an important input on empirical data about innovation networks: the case study on Virtual Centres of Excellence (VCE) analyses the nature and effectiveness of deliberately inaugurated innovation networks as a policy measure; the case study on biotechnology is about new products (product innovation); the Knowledge-Intensive Business Services (KIBS) case study concentrates on the co-evolution of technology development and innovation networks in e-commerce (producer-user networks); and the case study on Combined Heat and Power (CHP) technology is mainly concerned with the role of specific structures within the environment – in this specific case a centralised service supply system – on the dynamics of innovation networks.

Virtual Centres of Excellence (VCE)

The subject of our case study, the Mobile VCE in the area of mobile and personal communications, is an example for policy induced networks. As a political construct this 'Virtual Centre of Excellence' was addressed to issues of research quality (excellence) and of increasing public welfare by R&D – the most important concerns in science policy today. It was set up in 1996, as a consortium involving seven UK universities and almost all the major European companies active in mobile communications with the help of UK government funds. But government assistance was not confined to the provision of funds, civil servants from the Department of Trade and Industry (DTI) actively facilitating the design of the requested research proposals.

The term 'virtual' was used to assign 'geographically distributed', that is, bringing together existing small teams. The rationale of this construction addressed a perceived problem of the fragmentation of research in the UK. The suggested solution, a virtual centre, was a way of bringing the fragments together without actually moving researchers from their institutions. This should avoid the financial and other costs of a more rigid institutionalisation.

The aim of the virtual research network in Mobile VCE was to encourage inter-institutional work among research associates, as well as to create links between researchers and the industrial representatives involved in research management activities. The political goals of creating a virtual centre included the creation of a centre of excellence (by bringing together isolated research teams); and the encouragement of industrial-academic collaboration through providing industry with access to a wide range of the best academic work in relevant fields.

> The regulatory framework and management structures set up for Mobile VCE provide a context in which the various actors associated with the organisation can relate to each other through both formal and informal links, facilitating 'networking' in more than one sense. The actors in Mobile VCE are primarily its academic and corporate members (through their various individual representatives) plus the executive director and government observers. Relations between the actors are partly mediated by the VCE's formal management structures, particularly through meetings of the Steering Committees of each of the Core Programme areas (see Chapter 2).

While Mobile VCE undoubtedly 'worked' according to the terms in which it was set up, this does not necessarily mean that VCEs are, in general, a useful policy tool. In comparison, another VCE for digital television technology, the Digital VCE, was very much less successful. Some of the specific issues relating to innovation networks as a policy tool can be discussed in this case study. We have found two sorts of alliance supported by Mobile VCE which are specific (though not necessarily unique) to VCEs as a funding mechanism and are of interest as a policy tool: the virtual links of the research network and the self-perceived identity of the industrial network. The virtual research network in Mobile VCE encouraged inter-institutional work among research associates, as well as creating links between researchers and the industrial representatives involved in research management activities. The political aims of creating a virtual centre included the creation of a centre of excellence (by bringing together isolated research teams) and the encouragement of industrial-academic collaboration by providing industry with access to a wide range of the best academic work in relevant fields. Both these aims are better served by a VCE than by other policy initiatives aimed at forming research consortia. A VCE, it might be said, brings together two networks (the academic and the industrial), not merely individual actors.

The Case of Biotechnology

Biotechnology is not an industrial sector but a scientific field underlying a number of industrial sectors (pharmaceutical, agriculture, food, environment, etc.), here called the biotechnology-based sectors. Industrial applications of biotechnology are highly dependent on new scientific developments, even on those that are the result of basic research. The time between the creation of new knowledge and the funding of industrial research aimed at its applications is in general very short. Basic research is not exclusively confined to public research institutions, but is also carried out by firms. Thus, both for what concerns its intensity of knowledge utilisation and for the mechanisms employed, biotechnology seems to be a very good example of innovation dynamics in the context of science-based industries. However, a very high rate of growth of knowledge production involves a very heavy, irreversible and risky commitment. The collaboration with research institutes within molecular biology constitutes a more flexible and reversible strategy.

The case of biotechnology is of particular interest since some innovation networks have emerged spontaneously, while others have been created deliberately as a result of government policy. In the latter case, the network establishment itself is part of the objectives of the participants in the programme. In other words, the network is reflexive about the need to create itself, and for this reason the network formation follows a dynamic that differs from spontaneous network formation. Therefore, differences in national policies might explain different network structures in European countries. In a comparative study of networks on biotechnology, Saviotti (1996) found that French innovation networks were often politically motivated, while British networks emerged as a result of a convergence of individual interests. At the same time, the co-ordination structure was more centralised in French networks, while British networks operate in a more heterogeneous way.

Knowledge-Intensive Business Services (KIBS)

Knowledge-Intensive Business Services (KIBS) providers are private sector organisations that rely on professional knowledge or expertise relating to a specific technical or functional domain. KIBS can be primary sources of information and knowledge (through reports, training, consultancy) or providers of services to form key intermediate inputs in the products or production processes of other businesses (for example, communication and computer services).

An important feature that distinguishes KIBS from manufacturing firms is the type of 'product' they supply. Whereas manufactured products and processes contain a high degree of codified knowledge (they are a 'commodification' of knowledge), KIBS products contain a high degree of tacit ('intangible') knowledge. KIBS thus acquire special significance as agents who transfer experience and technologies within, and across, innovation networks.

As well as being vehicles of knowledge transfer, KIBS are engaged in the co-production of *new* knowledge and material artefacts with their business clients. This interactive problem solving is the 'product' that clients wish to purchase. Given the importance of this interaction, the factors that facilitate successful KIBS-client interaction should not be overlooked. The quality of the provider-client interaction depends on the competence of the clients as well as the KIBS supplier. Therefore the development of KIBS activities involves learning through networking, rather than networking alone. Consequently, innovation networks which involve KIBS activities are high in structural complexity.

Members of the networks are: users (businesses and households connected to the internet and interacting with it); communication lines and communications equipment providers; intermediaries (the suppliers of on-line information or access providers); hardware manufacturers; software authors and manufacturers (for example, browsers, site development tools, specific applications, smart agents, search engines and others); content producers and providers (for example, media companies); suppliers of financial capital; and public sector institutions.

Combined Heat and Power (CHP)

This case study is not only on innovation but also on diffusion of a new technology. Although co-generation, that is Combined Heat and Power (CHP), is not a very innovative technical principle, it has undergone substantial changes in the last 15 years. These changes combine technical improvements with substantial organisational innovations. They have been paralleled by major reforms of the energy supply systems in many European countries, which are going to be further modified in the next years within the new framework of the Internal European Energy Market. Therefore, policy networks are taken explicitly into account as complementary to innovation networks. Both types of network are tied together and supported by a third – information networks. These three are regarded as networks that underlie and drive the transformation of large socio-technical systems (LSTS) such as energy supply.

Due to these wide contextual transformations, the networks which have been established around CHP pervade many areas of society. The networks of technology suppliers (universities, R&D departments, manufacturers, system providers) and technology users (service providers, utility companies, energy end-users) play an equally important role as those which are concerned with regulatory reforms and energy technology policies.

Consequently, the structural complexity of innovation processes and networks regarding CHP is quite high. Characteristic of its dynamic is the mutual shaping and co-evolution of the innovation, policy and information networks on the one hand and of their structural or system context within the large socio-technical system of energy supply on the other. These interdependencies between networks and their systemic structural context are crucial in order to be able to take diffusion aspects into account. Not only do properties of the network emerge, but also the structural context of the network is reshaped in the course of the evolution of the network and the innovation diffusion of the underlying technology. While many of the structural features of the system may be independent of the evolution of the network during the innovation phase, they are clearly affected once a wider diffusion sets in.

Innovations often require contextual adjustments to become successful and there are several different channels through which such contexts are shaped and transformed. In turn, context conditions constrain the innovations that can emerge and diffuse. Setting up innovation networks alone is often not enough to establish an innovation. There is a high risk that the innovative effort will stop at an experimental stage and will never reach the stage of a wider application. In many cases, this is due to a lack of 'embedding' of the innovations in a compatible structural context.

The CHP case also shows how important organisational innovations are in complementing the technical dimension of innovation. The existence of potential carrier organisations with an intrinsic interest in the innovation in question turned out to be a decisive factor for establishing innovation networks. In the Netherlands, a clear strategy in favour of CHP was pursued, which explicitly aimed at setting up and co-ordinating a network of actors for the advancement of CHP. A counter-example can be found in the UK where the establishment of innovation networks was left to individual companies, supported only by some limited measures for facilitating information exchange. This is reflected in a highly liberalised new framework for energy supply. The situation in Germany is less clear because elements of explicit support and co-ordination of CHP at different levels of government are combined with a strategy of delegating responsibilities for setting up innovation networks to the energy supply industries. The above mentioned decentralised decision structures also imply a decentralised notion of reflexivity about innovation networks for CHP.

THE SIMULATION MODEL

The tremendous development of and easy access to computational power within the last 30 years has led to the widespread use of numerical approaches in almost all scientific disciplines. Nevertheless, while for example the engineering sciences focused on the applied use of simulation techniques from the very beginning, in the social sciences most of the early examples of numerical approaches were purely theoretical. There are two reasons for this. First, since the middle of the twentieth century, starting with economics, equilibrium-oriented analytical techniques flourished and were developed to a highly sophisticated level. This led to the widely shared view that within the elegant and formal framework of linear analysis offered by neoclassical economics, the social sciences could reach a level of accuracy not previously thought to be possible.

Second, within the same period, new phenomena of structural change and transformation from the Fordist System to a knowledge-based society exerted a strong influence on the social and economic realms. Despite the mainstream neoclassical successes in shifting the social sciences to a more mathematical foundation, an increasing dissatisfaction with this approach emerged. For example, by the 1970s the benchmark of atomistic competition in neoclassical economics had already been replaced by the idea of monopolistic and oligopolistic structures under the heading of workable competition (for example, Scherer and Ross 1990). A similar development emphasising positive feedback effects and increasing returns to scale caused by innovation led to the attribute 'new' in macroeconomic growth theory in the 1980s (Romer 1990).

In addition to these stepwise renewals of mainstream methodology, an increasingly larger group claimed that the general toolbox of economic theory, emphasising rational behaviour and equilibrium, is no longer suitable for the analysis of complex social and economic changes. In a speech at the International Conference on Complex Systems organised by the New England Complex Systems Institute in 2000, Kenneth Arrow stated that until the 1980s the 'sea of truth' in economics lay in simplicity, whereas since then it has become recognised that 'the sea of truth lies in complexity'. Adequate tools have therefore to include the heterogeneous composition of agents (for example, Saviotti 1996), the possibility of multi-level feedback effects (for example, Cantner and Pyka 1998) and a realistic representation of dynamic processes in historical time (for example, Arthur 1989). These requirements are congruent with the possibilities offered by simulation approaches. It is not surprising that within economics the first numerical exercises were within evolutionary economics, where phenomena of qualitative change and development are at the forefront of the research programme.

The first generation simulation models were highly stylised and did not focus on empirical phenomena. Instead, they were designed to analyse the logic of dynamic economic and social processes, exploring the possibilities of complex systems behaviour. However, since the end of the 1990s, more and more specific simulation models that aim at particular empirically observed phenomena have been developed. Modellers have had to wrestle with an unavoidable trade-off between the demands of a general theoretical approach and the descriptive accuracy required to model a particular phenomenon. In this respect, the simulation approach could also be labelled a *policy-friendly-modelling approach* because it allows the implementation of specific empirically observable cases as well as the experimental analysis of the impact of alternative policy instruments.

CONCLUSION

Following the arguments summarised in the above sections as well as the empirical facts given in our case studies one can see that knowledge production can usefully be conceptualised in terms of innovation networks. Nevertheless, several basic questions are not answered. There is no clear definition of what an innovation network is. Rather, there are numerous models, each emphasising different aspects depending on the research questions of the model builder. It is also not clear whether there are common characteristics applicable to all spheres of innovation, or disparate phenomena with little or no commonality. Do the innovation networks in biotechnology have the same characteristics as those in telecommunications? Is it useful to treat the dynamics of knowledge production in the different sectors of our case studies as similar? Not very much can be found in the literature about the dynamics of innovation networks: how they arise, the growth processes they undergo, and the way they die or merge into other networks.

Therefore, it is necessary to elaborate a theory of what constitutes an innovation network, how it operates and what kind of dynamics it undergoes. Our simulation approach to innovation networks faces this challenge. The role of this approach is not to create a facsimile of any particular innovation network that could be used for prediction, but to use the method of simulation to assist in the exploration of the consequences of various assumptions and initial conditions; that is, to use the method of simulation as a tool for the refinement of theory. This follows Axelrod's (1997) description of the value of simulation:

> Simulation is a third way of doing science. Like deduction, it starts with a set of explicit assumptions. But unlike deduction, it does not prove theorems. Instead a simulation generates data that can be analyzed inductively. Unlike typical induction, however, the simulated data comes from a rigorously specified set of rules rather than direct measurement of the real world. While induction can be used to find patterns in data, and deduction can be used to find consequences of assumptions, simulation modeling can be used to aid intuition (Axelrod 1997, pp. 24–5).

As a simulation technique a multi-agent model is used; that is, each of the actors relevant for an innovation process is represented by an agent or 'object' in the simulation. The agents are designed to have the attributes of 'intelligent agents': autonomy, ability to interact with other agents; reactivity to signals from the environment; and pro-activity to engage in goal-directed behaviour. To model the knowledge that the actors possess, each is a given a kene, a structured collection of technological, political, social and economic capabilities. A kene is used to represent the knowledge base of an actor. Kenes change as actors acquire knowledge from other actors, and as they refine their knowledge through research and development. Kenes are made up of capabilities, and each actor has one or more abilities for each capability. For example, a biotechnology firm might have the capability to synthesise a particular pharmaceutical ingredient using a specific manufacturing operation (its ability). Actors use the knowledge represented in their kenes to produce artefacts, which, depending on the setting, might be a new design, a new drug, an invention for which a patent application could be made or a new discovery publishable in the scientific literature.

These artefacts are merely potential innovations. Only a small proportion of them become innovations, that is, successful new products and processes. The selection of which artefacts are innovations is modelled by the Innovation Oracle. The oracle rejects unsuccessful artefacts and rewards actors who produce successful innovations. For the purpose of the model, the oracle maintains a multi-dimensional 'innovation landscape' onto which all possible artefacts can be mapped. The 'height' of the landscape at the point where the proposed innovation is located is used to determine whether the artefact is an innovation, and if so, the amount of reward that flows back to the innovating actor. The form of the landscape is complex and unknown to the actors and they cannot anticipate with certainty how successful their innovations will prove to be. For this reason the model does not specify a specific innovation but the possible reward of a successful innovation. Novelties deform the landscape so that the reward for a similar artefact identical to the first innovation is much reduced (this models the fact that in most fields the first mover will patent or copyright their innovation, with the result that subsequent identical artefacts from other producers receive little or no reward). As a result, the landscape co-evolves with the actors' kenes, and this is one way in which the model reproduces the complexity of the real world.

An actor's kene can develop as a result of three factors. First, an actor can use its resources to engage in 'incremental' or normal research and development (R&D). This improves the actor's ability relating to a specific capability, using its experience with previous artefacts. Second, the actor can elect to engage in 'radical' research in which entirely new artefacts are created from new combinations of the agent's capabilities. This corresponds to a firm deciding that it needs to branch out into a new area of expertise. Third, an actor can learn from another actor with whom it is collaborating in a partnership. Partners share their knowledge when a partnership is formed, thereafter producing their own artefacts from their own kenes.

Actors are always available to join partnerships and whether they in fact do so depends on their strategy for developing their kenes, and also on whether they are able to find partners that are sufficiently attractive. All actors display an 'advertisement' consisting of a list of the capabilities that they possess (but not the details of the actual abilities that they have, since these will be confidential and in many cases, are not easily made explicit). For example, a biotechnology start-up firm might advertise that it is capable of performing some aspect of genomics, thus making itself attractive to other firms without that expertise and possibly to venture capitalists or large firms seeking partners to develop new markets. Actors use the advertisements to judge whether the capabilities of a potential partner are sufficiently tempting to warrant forming a partnership. The costs of networking include sharing the rewards of innovation and the fact that partners have to give away their own knowledge as well as acquiring that of others.

Partnerships are considered to be relatively short-term relationships focused on the development of particular products. They are typically binary relationships, although some actors may enter into a number of collaborations with different partners. In some fields, more permanent and more densely connected networks are also found. These networks always include a number of actors, bound into a collaboration which is more enduring than a partnership and which often has a distinct identity (ranging from an informal name to a legally based company, as with the Virtual centre of Excellence in Mobile and Personal Communications, which is a non-profit limited company with its members as the shareholders). In the model, agents are able to convert a set of partnerships into a network provided that all the members have previously been partners. The advantage is that the network pools its members' capabilities and has a lower cost of collaboration than would have been the case if the members had only binary partner relationships.

The actors are driven by a need to maximise their rewards. Set against the rewards provided by the oracle for successful innovations are the costs of performing R&D and of collaboration and networking. Actors that fail to accumulate sufficient rewards to keep themselves in funds are declared bankrupt and 'die'. On the other hand, if the population of actors is successful in earning rewards, start-ups enter the field, copying the kenes of the most successful actors and thus increasing the level of competition.

To summarise: actors as the basic elements of an innovation network are defined by their kenes, their competencies, and their structural and behavioural attributes which determine their ability to observe, anticipate and design the product space. Because these characteristics differ for each actor, innovation networks are heterogeneous. The main problem of an innovation network is to co-ordinate these heterogeneous actors. The co-ordination mechanism is the mechanism of self-organisation: due to the increasing complexity of innovations so called 'go-it-alone' strategies are becoming more and more risky and co-operation with other actors is one way to reduce the uncertainty of the innovation problem. The emergence of innovation networks is driven by this uncertainty. The participation in an innovation network allows actors to have access to the capabilities of other actors which are not available by other means because of their tacit or local components. However, choosing a co-operative strategy also means sharing one's own knowledge with others. This is also a source of uncertainty because partners in an innovation network may become competitors in the market when the new product is developed. Balancing the risks and the advantages of co-operation is the main mechanism of social integration within an innovation network. This is a process of self-organisation because there is no external reference by means of which to find the appropriate way.

An innovation network produces an innovation. Depending on the context this might be a new design for a car, a new drug, new knowledge for which a patent is possible, or a new effect which could be used for a new technology. The production of these 'products' is also a self-organised process: the knowledge which is needed for the new product is not available, it must be produced and the production of knowledge can be characterised as a selection process applied on different hypotheses which have to be verified on the basis of its practical consequences.[2] Within the simulation model this selection procedure is conceptualised as an 'Innovation Oracle'. This is a kind of black box which evaluates and selects those 'proposals' that are to count as innovations. If the Innovation Oracle has passed the proposal, it has proved to be a successful innovation and a reward is paid to the successful innovators. If the Innovation Oracle refuses a proposal the innovation is not seen as being successful. Then the information oracle generates information about the type of knowledge that might be required to improve the current approach. The Innovation Oracle serves as a selecting environment for possible innovations.

The computer simulation model of innovation networks helps to support, communicate and legitimise not only strategic decision within the planning of a concrete innovation process. At the same time it helps policy decision processes. In a way, applying IT techniques to the policy process is a self-referential operation which follows the rationale of technology policy itself, which is namely to strengthen the linkages between science and users. Here policy makers can apply funded technology to their own area. This feedback or even payback between science policy and IT simulation research is the same mechanism observed within the self-organisation of innovation networks. Therefore such a simulation model can also be used within Research Technology Development (RTD) policy especially in evaluation processes.

Because innovation networks integrate structural and dynamic perspectives, a radical change in evaluation theories, methods and tools is required. It is not only the product and its impact which is the focus of evaluation, it is also the production within a network which requires new concepts of evaluation. Problems of the performance of the network, the integration of the context of application, the problem of social accountability pose new questions for evaluation. Chapter 9 on new evaluation techniques shows how the simulation model can be used for this purpose.

NOTES

1. Measures of structural complexity are quantities such as entropy, redundancy, information and the like.
2. In the case of scientific hypotheses and experimental data see, for instance, Krohn and Küppers (1989).

REFERENCES

Arthur, W.B. (1989), 'Competing technologies, increasing returns and lock-in by historical events', *Economic Journal*, **99**, 116–31.

Axelrod, R. (1997), 'Advancing the art of simulation in the social sciences', in R. Conte, R. Hegselmann and P. Terna (eds), *Simulating Social Phenomena*, Berlin: Springer-Verlag, pp. 21–40.

Bell, D. (1973), *The coming of Post-Industrial Society: A Venture in Social Forecasting*, New York: Basic Books.

Cantner, U. and A. Pyka (1998), 'Absorbing technological spillovers: Simulations in an evolutionary framework', *Industrial and Corporate Change*, **7**, 369–97.

Freeman, C. (1991), 'Network of innovators: A sysnthesis of research issues', *Research Policy*, **20**, 499–514.

Gibbons M., C. Limoges, H. Nowotny, S. Schwartzman, P. Scott and M. Trow (1994), *The New Production of Knowledge: The Dynamics of Science and Research in Contemporary Societies,* London: Sage Publications.

Henderson, R. and K. Clark (1990), 'Architectural innovation: the reconfiguration of existing product technologies and the failure of established firms', *Administrative Science Quarterly*, **35** (March 1990), 9–30.

Krohn, W. and G. Küppers (1989), *Die Selbstorganisation der Wissenschaft,* Frankfurt/M.: Suhrkamp.

Küppers, G. (ed.) (1996), *Chaos und Ordnung: Formen der Selbstorganisation in Natur und Gesellschaft*, Ludwigsburg: Reclam.

Küppers, G. (1999), 'Coping with uncertainty – New forms of knowledge production', in K.S. Gill (ed.), *AI & Society*, London: Springer-Verlag, pp. 52–62.

Machlup, F. (1984), *Economic Information and Human Capital,* Princeton: Princeton University Press.
Malerba, F. (1992), 'Learning by firms and incremental technical change', *Economic Journal*, **102**, 845–9.
Pyka, A. and P.P. Saviotti (2000), *Innovation Networks in the Biotechnology-Based Sectors* (SEIN-Project Paper No. 7,
online: http://www.uni-bielefeld.de/iwt/sein/papers.html).
Romer, P.M. (1990), 'Endogenous technical change', *Journal of Political Economy*, **98**, 77–102.
Rycroft, R.W. and D.E. Kash (1999), *The Complexity Challenge. Technological Innovation for the 21st Century*, London and New York: Pinter.
Saviotti, P. (1996), *Technological Evolution, Variety and the Economy*, Cheltenham, UK, and Lyme, US: Edward Elgar.
Scherer, F.M. and D. Ross (1990), *Industrial Market Structure and Economic Performance*, Boston: Houghton Mifflin Company.
Stehr, N. (1994), *Arbeit, Eigentum und Wissen, Zur Theorie von Wissensgesellschaften*, Frankfurt/M.: Suhrkamp.
Zuscovitch, E. and M. Justman (1995), 'Networks, sustainable differentiation, and economic development', in D. Batten, J. Casti and R. Thord (eds), *Networks in Action, Economics and Human Knowledge*, Berlin: Springer-Verlag, pp. 269–86.

2. Complexity, Self-Organisation and Innovation Networks: A New Theoretical Approach

Günter Küppers

THE PROBLEM

Innovation theories have regained their popularity. Returning once more to Schumpeter, who first made economists aware of the role of innovations for economic change, the dynamics and structure of innovation processes have become a focus of interest again, not least because innovation has gained new importance as a key term in sociological modernisation theories. For some time now, the planning, production and marketing of new products and procedures has been perceived as a problem. This problem has two decisive dimensions. Empirical findings indicate that the innovation pressure is expanding continuously because:

- Products are developed increasingly for potential customers and their application context. One can see a transition from the production of unspecific mass goods to the production of special products designed for specific tasks.
- The broader and more rapid deployment of knowledge through new information technologies and the growing importance of knowledge for the dynamics of innovation increase the competitive pressure and, thus, also that of innovation. New products have to be launched on the market more and more quickly if they are to be an economic success.

Parallel to this, innovation decisions become increasingly more risky because:

- Innovations are growing in complexity.

- The market is becoming increasingly less transparent and more turbulent, and launching a new product is becoming more of a risk. For a new product, quality alone is no longer decisive for its commercial success.
- The third reason for the increasing risk of innovation can be seen in a growing dependence of the success of an innovation on the available knowledge. Labels such as the 'knowledge society' and 'knowledge-based industry' point to a fundamental change in the role of knowledge. It has now become the raw material for technological and social innovation processes.

From the current theoretical perspective, innovation is neither an adaptation to perceived needs nor a unique act of (linear) transformation of inventions into products.[1] In particular, the (more meso- and macro-structurally oriented) evolutionary economics, but also the (more micro-structurally oriented) sociologically oriented technology studies rejected the causality assumptions of classic innovation theories in their theoretical and empirical studies and are developing models describing innovation processes as being multi-referential, non-linear, dependent on various framing conditions and rarely predictable.

Continuations of these analyses have managed to show that innovations are produced regularly in networked structures ranging from producer-user relations to complex networks of science, business, politics, law and administration. The common concern of network theories in innovation research is to bring about a shift in perspective from the analysis of industrial innovation processes to the organisation of innovation processes among and between companies, and thereby replace mono-causal explanatory schemes (economic, technological, power-theoretical, etc.) with more complex assumptions. Modern innovations are viewed neither as the outcome of invention and marketing strategies in individual companies in response to complete knowledge about relevant demand states, nor as fitting the picture of a passive transformation of scientific knowledge into applied knowledge and then – mediated by company R&D departments – into products that are ready for the market. In contrast to the traditional perspective, modern innovation research has created a far more differentiated picture of technology development in which the concepts of uncertainty, lack of transparency and acceptance play a central role.

Nonetheless, the theoretical description of innovation networks remains controversial. These controversies address their demarcation from the role of environment and other forms of organisation, ways of functioning and negotiating within networks, the meaning of contracts and trust, the mechanisms of conflict regulation and the importance of learning. Further research issues refer to the management of cognitive and social (organisational) uncertainty and the momentum of innovation networks. The purpose of this chapter is to make a substantial contribution to this theoretical discussion. It starts with a presentation of the state of the art. Two major approaches will be discussed at the beginning: the contributions of evolutionary economics and the sociological analysis of technology production. In the following an integrated perspective of analysis is developed which is based on the concept of self-organisation. Before this concept is discussed in greater detail a section on complexity shows the meaning of this central term in the discussion about modern innovation processes, a term which often is used without explaining its main characteristics.

THE STATE OF THE ART

Evolutionary Economics[*]

Since the beginning of the 1980s, a new direction has developed in research within the conflict between the classic theory of innovation and Schumpeter's approach: evolutionary economics.[2]

This new departure in research has attracted a remarkable amount of public interest within a short time. It has had its own journal since 1990 (*Journal of Evolutionary Economics*) and, since the beginning of the 1990s, its own network (EUNETIC = European Network on the Economics of Technological and Institutional Change) of research groups funded by the European Union.[3]

The (neo-)Schumpeterian outlook on innovation through evolutionary economics proceeds from a radical criticism of the (neo-)classical theory of technological change. This criticism addresses the theoretical modelling of innovation dynamics (Dosi 1983), both the theoretical and methodological instruments as well as the conclusions of empirical studies (Rosenberg 1982), and the actor-related implicit assumptions on rationality (see Silverberg 1988; Nelson 1995).

The term 'evolutionary' covers the dynamic perspective on economic processes (dynamics first) and the central assumption that the evolutionary explanation of innovation processes must integrate not only the randomly determined variation of elements in the innovation process but also their selection (see Dosi and Nelson 1994, pp. 154f.; Marengo and Willinger 1997, p. 332). 'Evolution' implies the interaction of principles of variation and selection and presupposes the heterogeneity of the elements on which both principles impact.

---

[*] Co-authored with Uli Kowol

In fact, evolution implies the interaction of a principle of variation and a principle of selection. The former introduces into the system two crucial features at the micro-level: novelty and heterogeneity. The selection principle is meaningful only if there is at least some heterogeneity among the micro-units on which it operates; without heterogeneity there is nothing to be selected for or against and there is no scope for evolution. Moreover, without the introduction of novelty at the micro-level such heterogeneity cannot be sustained in the long run and evolution comes to an end (Marengo and Willinger 1997, p. 332).

Nonetheless, evolutionary economics is not a coherent theory with a common methodology. It is more of a juxtaposition of various approaches sharing certain features: the critique of the neoclassical approach, the rejection of the assumption of rational decision making and support for the examination of dynamic processes within the economy. Whereas some proponents of this trend place more emphasis on empirical work and have developed taxonomies (radical vs. incremental innovation, high-tech vs. low-tech firms, small vs. large companies, etc.), others favour (institutionalist) approaches in which the dynamic of innovation is explained in terms of the complex interplay of individual commercial innovation decisions, networks between companies and institutional settings (see Powell and DiMaggio 1991; Dosi and Nelson 1994). What they all emphasise is, first, that technological change is under-determined by technological factors, and, second, that successful innovations do not always have to be the best ones, but represent the outcome of multiple and contingent effective factors. In this theory programme, technological change is not determined exclusively by either the economy or technology. The openness to design of new technologies is faced on the one hand with closure processes and lock-in effects on the other. Both organisational and institutional aspects as well as learning processes on different levels of organisation and society play an important role in formulating the theory of evolutionary economics (see McKelvey 1996).

The theoretical foundations of evolutionary economics also draw on different sources: the long tradition of formulating theories in economics on the basis of mechanics has shifted to a stronger reliance on biological models. Irreversibility in time and the consideration of dynamic processes come to the fore (Dosi and Metcalfe 1991). The assumption of complete rationality in actors is replaced by theoretical formulations based on game theory and institutionalism. In recent times, computer simulations that are based on biological evolutionary theory and work with evolutionary or genetic algorithms have also been used (see Birchenhall, Kastrinos and Metcalfe 1997; Brenner 1998; David and Kopel 1998).

In a (mostly) broad analogy to evolutionary theory, the following analytical and methodological extensions can currently be recognised in the research programme of evolutionary economics:

Analogously to biological evolutionary theory, recent evolutionary economics uses the concept of 'fitness' (Dosi and Nelson 1994) to emphasise the social construction of technology rather than a technological determinism, and thus point to the potential for variation. Accordingly, technological solutions have only to fit or be suitable for an application context, and not achieve an optimum.

The selection criteria developed originally for evolutionary economics ('profit' in Nelson and Winter 1982; 'prices' in Silverberg 1988) are currently giving way to a more complexly constructed selection dynamic. This posits that companies select according to a variety of criteria (capital markets, anticipated profits, market development, economic growth conditions, product strength, prices, distribution conditions, etc.). This multidimensionality of selection criteria challenges evolutionary economics to specify the interaction mechanisms through which selection is performed. A further challenge is associated with the above-mentioned extension of the approach through organisational sociology: selection can no longer be located within an unstructured business environment. In the real world, users are not exclusively selectors but also involved in the shaping of innovations (see Anderson and Tushman 1990; Tushman and Rosenkopf 1992; for an overview, Kämper 1995). As a result, some authors conceive selection as an internal process within the 'community of producer and user' (Anderson 1991) or within a 'societal sector' (Scott and Meyer 1991). Other authors discriminate between internal and external selection. This follows, first of all, from the realisation that innovating companies do not generate innovations exclusively internally, and, second, that no innovation (no product) is purchased (selected) without being tested first (see Birchenhall, Kastrinos and Metcalfe 1997, p. 379).

An evolutionary model of technological change has to name not only the mechanism of variation and selection but also a mechanism responsible for the stabilisation of a successful variation (see, also, Campbell 1965). Anderson and Tushman (1990) see this in the formation of a 'dominant design'.[4] This phase of innovation dynamics is characterised by the formation of standards and technological norms (Anderson and Tushman 1990, pp. 613ff.). A dominant design initiates further standardisations of individual parts and – with a view to the manufacturer of technology – the optimisation of the assembly processes and the stabilisation of manufacturer-supplier relations, distribution networks and customer relations. From the perspective of the technology user, a dominant design reduces the lack of transparency over the variety of products in a given class, and reduces the costs of acquisition. Learning by doing and learning by using (see Rosenberg 1982) stabilise this design. User experiences lead to an improved understanding of mistakes and permit more reliability through better service. In terms of evolution theory, the successful stabilisation of a technology also changes the selected environment. A technological paradigm now structures further research activities and user needs. It can be suppressed only with great difficulty. This is also why there are dangers of a lock-in effect. For this reason, recent innovation theory also points to further necessary learning strategies for the innovating company, so that it is also equipped to deal with rapid technological change in its environment. Such learning strategies concern the levels of the organisation (organisational learning) and the inter-company, inter-organisational levels (recursive learning processes, see Krohn 1997a). As we shall see below, this extension has important ties with the sociological approach of technology studies and network theory.

On the basis of the 'evolution metaphor', evolutionary economics has recently developed a variety of analytical and methodological refinements to the design of the theory.[5] Evolutionary or genetic algorithms play a particularly important role here. The idea behind the instrument of genetic algorithms, borrowed from computer simulation (for a basic discussion, see Goldberg 1989; Holland [1975] 1992), is to develop optimisation procedures whose (modified) functions correspond to those of natural evolution. In particular, computer simulations are intended to model the interplay between variation, mutation and selection. In addition, the importance of learning is seen and models of learning through imitation are developed. To improve the modelling of imitation learning within a population, some authors replace the selection operator with another operator that they call 'selective transfer'.[6] They then relate this modified genetic algorithm to the selection of technologies in order to construct a model of technological change (see Birchenhall, Kastrinos and Metcalfe 1997, pp. 388ff). The significant aspect is that the discrimination between the unit of selection and the selecting environment found in biological evolution is abandoned in favour of a discrimination between internal and external selection (ibid., p. 379).

The Sociological Analysis of Technology

Whereas the main contribution of evolutionary economics has been to improve our understanding of macro- and meso-structural developments, the micro-sociological analysis of individual innovation processes delivers knowledge of the structure and dynamic of the relevant social interaction processes. It reveals that actor configurations, the poly-contextuality of the decision-making criteria and the contingency of the decision-making process are characteristics of the innovation process and not, as often claimed in classic, more historically oriented technology studies, scientific progress and technological rationality.

Beginning in the late 1970s, detailed analyses of the 'context' of technological innovations, that is, of the networks and boundary conditions around them, have provided rich empirical information (Bijker, Hughes and Pinch 1987; Callon and Law 1989). Unfortunately, the case studies were presented mainly as isolated narratives which were not integrated within a structured theoretical framework and could therefore not be constructively used in policy analysis. Systematic efforts under headings such as 'actor networks' (Callon 1992) or 'socio-technical constituencies' (Molina 1999) have provided only very rough and static patterns of analysis, but do not allow one to study and understand the dynamic behaviour of innovation networks. In addition, they assigned technological driving forces a very limited, if not negligible, significance in shaping technological innovations. The study of the evolution and transformation of large technical systems such as energy supply, telecommunications or transport, has shown that technological interdependencies and 'momentum' can in fact be strong determinants and constraints of future technological pathways (Hughes 1983; Hughes and Mayntz 1988; Summerton 1994).

Current research on the emergence of new technologies is focusing on the conditions that lead to the concrete form of a technology. Among other aspects, it is asking which visions of use and guiding principles are pursued or ruled out, whether and how these are reinterpreted and transformed during the course of developing a technology, how controversies between different functional demands and efficiency criteria are resolved, which micro-political constellations are decisive in shaping a technology, and which organisational and institutional conditions exert an influence on this (see Rammert 1993, 1994; Rammert and Bechmann 1994; Halfmann 1995). It has shown that assumptions regarding the possibility of controlling and planning technological change are inappropriate. Deterministic assumptions about one single best means of technological progress are successfully rejected and replaced by the social construction of technology.[7]

Although directed initially only toward generating new technologies, the application context has been discovered to be a determining factor and has been increasingly taken into account within this field of research since the beginning of the 1990s. Through the further consideration of institutional and organisational parameters within research on technology production, attention has also spread to the constitutional conditions and functional mechanisms of innovation processes. Although disagreements continue on how the theoretical concepts derived from the numerous single case studies – among others, studies are available in mechanical engineering (Kowol and Krohn 1997), the waste-disposal industry (Herbold, Krohn and Timmermeister

2000), computer industry and software development (Konrad and Paul 1999) – may be integrated to form a general model of technological change, the micro-sociological fine-scale studies of technological development provide important potential starting points for this venture.

It can be stated that technological development is not a linear-sequential process that can be traced back to rational decisions. Within the framework of research on technology production, the innovation concept describes the construction of an application context through a recursive process of transforming scientific and technological basic knowledge and anticipated visions of use, of institutional decisions and cultural models, practical experiences with the development of a technology, the commercial, market-oriented considerations, and, not least, the transformation of interests and utopias in order to shape the future into new marketable products and procedures that, in turn, (may) trigger numerous learning processes and feedback permeating as far back as the basic science.

As future-related projects, innovations processes are always fraught with uncertainty and contingency. They contain assumptions about future applications and uses. They model a social-technological application context in which the technological ideas are developed and, later, the artefacts are implemented experimentally. Modern innovation processes couple the technology developers' visions of use with the applied practice of users. The observation of these social closure processes between the generation, application and regulation of innovations has led to a new way of modelling innovation dynamics: the model of innovation networks.

The figure of the recursive coupling of visions of use[8] (intentions) and applied practice (expectations) in the innovation process forms the starting point in modelling theory for the recent, network-like analyses of innovation dynamics. The various conceptions of the 'innovation in the net' (Rammert 1997) thus counter the biology-laden basic scheme of (technological) variation and (market) selection found in evolutionary economics with an alternative basic model. In the case of technological innovation, the starting point is the construction of a new technology while taking account of its application context. One may object that application contexts do not always need to be built completely from scratch – partly because they emerge by themselves as a result of a clear demand (innovation through the market) or because they are already pre-structured sufficiently by existing markets (innovation through organisation). Although this may (continue to) be true in individual cases, to the extent that the innovation dynamic is confronted with increasingly turbulent markets and is burdened with technological, arithmetical and application-specific uncertainties, the origins of new application context also need to be integrated into the theoretical concept (see Kowol and

Krohn 1997). White (1981) has expressed this point succinctly in his title 'Where do markets come from?'.

The Need for an Integrated Perspective

Neither evolutionary economics nor the different approaches within the research programme of social studies of technology are able to describe the emergence, structure, and dynamics of innovation networks in an appropriate way. Within the models of evolutionary economics, three important aspects have been neglected above all.

Although learning plays a major role in modern concepts of evolutionary economics, the latter would seem to suggest that it is problematic to apply it to social development (Brenner 1998, p. 281). There are no statements on the motivation underlying learning processes. Nothing is said about the different ability of individuals to learn. The application of the concept of fitness may be correct in a biological context, but possesses insufficient complexity for social processes in which each individual within the social environment may well understand the term in a different way (ibid.). When formulating a theory of innovation dynamics, it would appear to be meaningful to abstract from biological evolution and seek a new reference point. This also seems worthwhile because modern biology and genetics have passed beyond the foundations of Darwinism, and there is a broad acceptance of concepts of self-organisation (see, for an overview, Foster 2000, pp. 313ff).

As long as revolutionary changes are disregarded, new technologies (innovations) can certainly be viewed as variations of existing ones. However, they do not vary blindly, because they have an application context that has already been anticipated in the draft of the construction.[9] Nonetheless, the impact of this application context is, primarily, not selective but constructive, and it has an accelerating impact through positive and negative feedback. Unlike evolutionary economics or biological evolution theory, cyclical couplings between variation and selection are assumed in the case of innovations. The most important point in modelling theory is that the structuring of need is a matter for the technological draft. It is precisely here that we find the feedback between technological variation and need variation. The reciprocal coupling between variation of needs and technological variation through which the model achieves its cyclicity is already available in any case in the sense of the demand model. Such a non-linear process dynamic also departs from the constraints of classic Darwinism, because its basic formula of blind or randomly determined variation coupled with a systematically effective selection is no longer applicable when there is no way to distinguish exactly between variation and selection (see also Krohn and Küppers 1989; Kowol 1998, p. 24).

Another point becomes important: innovation processes are dynamic. There must be a driving force responsible for a continuous development of new products. What drives the process that permanently allows the new to emerge? Foster (2000) proposes introducing competition as the driving force of innovation dynamics.[10]

Typically, the fear of competition is itself a stimulus for innovation, and thus the tendency towards a stationary state affects the stationary state itself and we have a non-equilibrium process (Foster 2000, p. 325). Competition can be seen as an uncertainty and new products are a measure to reduce this uncertainty. If they are cheaper, better, more reliable, etc., than the old ones, they decrease the pressure of competition and open up new possibilities for economic success. But the success of a specific product increases the innovation pressure for the others. Therefore, competition is ever-continuing and makes the economy a dynamic system. Within this dynamic, innovations are becoming more complex. The use of new technologies becomes important for the development of new, competing products and the question arises whether Schumpeter's person-centred image of the 'creative destroyer', the 'dynamic entrepreneur', can be retained in a world of complex innovations. Is it not far more the innovative business organisations that are responsible for permanent variation through co-operative exchange, knowledge accumulation and networked innovation activities with users, universities and research institutes?

The modern approach to innovation makes innovation networks responsible for reducing this complexity by subjecting the co-operative draft of an innovation to appropriate eligibility tests in various phases of recursive learning processes. The generation, application and regulation of new technologies are therefore not coupled in the sense of a trial-and-error procedure of variation and selection, but through a constructive knowledge production and learning process that is co-ordinated between the actors engaged with the aim of reducing the uncertainty accompanying complexity. Therefore, these knowledge production and learning processes are directed toward the management of uncertainty. The number of actors involved is determined by the uncertainty dimensions affected by the innovation: time frames, technological complexity, financing arrangements, knowledge deficits, the legal situation, risk perception, public sensitivities, and so forth. But new problems arise. Why should individual and collective actors enter into network-like relationships? The general answer is that individual actors foresee a potential gain through co-operation that cannot be attained through an individual strategy for maximising utility. The basic hypothesis is that alter controls resources (capital, manpower, knowledge, experience, etc.) that ego needs to increase utility and vice versa (Lundvall 1992; Kowol and Krohn 2000). The controversy is over whether the way co-operation is modelled in game theory (rational choice theories) provides an appropriate reflection of the functions and structure of networks (see Johnson 1992). If not, the question arises which theoretical alternatives do exist.

A second point concerns how a network becomes established: when opportunistic behaviour is possible, the forms of control are low, and thus the risk involved in co-operation is high for both partners, it has to be asked who will initiate the 'risky investment in co-operation', and how does a network come about? Few network theories offer an explicit explanation of the conditions under which networks emerge (see Rölle and Blättel-Mink 1998). Those approaches that address this issue in concrete terms generally proceed from the problem of reciprocal market and organisation failure (lack of transparency, arithmetic and cognitive uncertainties, turbulent organisation environments, etc.) and view the concept of trust as the central mechanism for forming a network (see Powell 1990; Mayntz 1992; Kowol and Krohn 1995). Nonetheless, this is not always followed by a specification of the concept of trust in sociological terms.

Nonetheless, it also has to be taken into account that co-ordination and co-operation based on trust do not always have to succeed. Problems not only with emerging threats of opportunism but also with knowledge transfer create blockades – and blockades resulting from professional cultures generate potential barriers that are not easy to manage. Finally, problems with divided loyalties may well prove to be the special Achilles' heel of innovation networks: network actors have to be loyal not only to their network but also to their focal organisations. The fragile structure of a network will be threatened when differences between the expectations and interests of the network and the focal organisations emerge as a problem of twofold commitment, and actors are confronted with contradictory demands. Finally, it should also be mentioned that innovation networks are able to generate externalities that have hardly ever been described theoretically up to now.

A third point concerns how far networks exist. The former position is supported particularly by Williamson's (1990) transaction costs theory. However, most network analyses follow Powell's (1990) within or beyond the market and the hierarchy of classical organisations proposal that defines networks as an autonomous third form of co-ordination alongside the market and the hierarchy (Freeman 1991; Teubner 1992; Grabher 1993; Kowol and Krohn 1995). What is discussed controversially here, when considering strategic business or supplier networks, is how far such emergent forms of co-operation can also be power based, in other words, hierarchical (see Sydow 1992; Pohlmann 1996; Sydow and Windeler 1997). In the case of innovation networks, discursive negotiation and co-operation based on trust are viewed as a central form of co-ordination, because networks are dominated by another type of social integration that deviates not only from formal contracts dominating co-ordination on the market but also from the command principle of a hierarchical organisation (Kowol and Krohn 1995).

Furthermore, the relations between actors within networks are described in terms of reciprocity, learning and preferential treatment. The literature on innovation theory had already emphasised the importance of learning at an early stage: Arrow (1962) referred to 'learning by doing', and Rosenberg (1982) added the role of 'learning by using' in the application context of new technologies.[11] Various concepts of organisational learning have been imported into innovation theory from organisational sociology (see Argyris and Schön 1978; Senge 1990; Wiesenthal 1995; Probst and Büchel 1998). Learning processes in the context of manufacturing and applying new technologies are viewed as a consequence of positive feedback (with corresponding acceleration effects). At the same time, they present the filtering out of negative feedback (with corresponding losses of experience). Network theory favours the concept of recursive learning for these reciprocal learning processes (see Krohn 1997a; van den Daele and Krohn 1998).

The conception of innovation networks models not only relations between manufacturers and users of a new technology but also between other actors. Alongside government actors, who (are supposed to) adopt a controlling, regulating and promoting function, increasing importance is assigned to the universities, non-industrial research organisations, standardisation committees, business associations and chambers of commerce, labour organisations and a variety of social movements.

Despite all the empirical evidence of the existence of innovation networks and their phenomenological descriptions it is still an open question how individual actors are going to co-operate within a new form of organisation which gains its own identity. In line with recent ideas in self-organisation theory, the process of networking heterogeneous (individual or collective) actors can be described as a process of dynamic structure formation. Such processes lead to a circular, autonomous form of operation that hardly ever tolerates external control. It is unlikely that a dominant (controlling) actor will run a network. Therefore the dynamics becomes complex, rules of co-operation emerge and external control must be replaced by self-control.

The approach of self-organising innovation networks aims to answer these questions by proposing an integrated theoretical framework. It takes into account concepts of self-organisation that reflect the parallel operation of a great variety of forces from formerly separate social spheres on the innovation process and provides a better understanding of knowledge production and innovation. It covers micro-aspects (such as actor strategies and networks) but also the context and structural conditions of innovation dynamics. In the following paragraphs of this chapter an overview is given of the theoretical concepts used within this approach – complexity, systems, self-organisation and innovation networks.

Complexity: Between Certainty and Uncertainty

Unfortunately, there is no general definition of what is complex as opposed to complicated. As a rule, structures are defined as complex when the sequence of elements from which they are made neither follows a regular pattern nor is determined at random. Therefore, neither the sequence of numbers 0101010101 nor the sequence of the numbers 0 and 1 produced at random through the toss of a coin (head = 1, tail = 0) is complex. Complexity in this sense represents the domain between the regularity of order and the irregularity of chaos. For the world of processes, this definition is of only limited value. However, a process can also be described as a sequence of events – the stock market, for example, as a sequence of share values at a fixed point in time or the weather as a sequence of temperatures at midday – and then the above-mentioned ideas on structural complexity can be applied to these number sequences.

However, this still does not disclose anything about the complexity of the dynamic producing these events. This dynamic, in turn, may be made up of a number of single processes. Such a linkage of single processes into a total process always occurs when individual elements, each with its own dynamic, merge into a whole and generate a total dynamic that differs from the individual one. This total dynamic may be a 'superposition' of each single process so that the total process can always be broken down into its sub-processes, or it may develop a new entity whose properties can no longer be read off from the single processes. This emergence of a new level, a new phenomenal domain, is a characteristic of dynamic complexity: the whole is more than the sum of its parts, or the whole has properties that are lacking in the parts. This was observed by Aristotle, who lists living things, souls, the state and works of art as examples of systemic wholes. In each example, the systemic properties do not result from adding together the properties of their components. Furthermore, the components of the system possess their individual functions only as parts of the whole. Hence, the system generates the functional properties of its elements and a separation into the individual components leads to the loss of the functions. This is why Aristotle argues that the whole is of necessity more original than the part (see Aristotle 1959, p. 28).

Here as well, complexity is linked to a loss of predictability. The dynamic of the whole cannot be predicted, because the whole has repercussions on the parts, and part and whole, element and system, are linked together in a dissoluble dynamic relationship. Both levels can no longer be viewed in isolation; the one level cannot be reduced to the other. One can neither see the contributions of the elements to the system, nor make sound statements about a specific element by analysing the system. This reciprocal conditional relationship leads to a type of 'uncertainty principle': system and element levels cannot be observed at the same time. When looking at the elements, the view of the system becomes blurred, and, vice versa, the view of the elements is blurred as long as one is focusing on the system.

The 'intensity' of the linkage between the whole and its parts, between the system and its elements, can be used as a measure of complexity. If it is non-linear, it ensures a degree of ambiguity (not arbitrariness) and as a consequence, a path dependence of the dynamic. Complex systems are therefore historical; they depend on their history. Whereas, under linear conditions, behaviour always remains constant, non-linearity ensures variations. In the linear, causes and effects are proportional to each other; which is why small causes always have small effects. In the world of the non-linear, effects are no longer necessarily proportional to their causes. Small causes may have large effects because they may amplify themselves through the non-linear effects. Small fluctuations in the form of internal fluctuation or external perturbations from the environment on the micro-level may amplify into large effects on the macro-level and drive the development of the whole in various directions. Complexity can lead to unpredicted (and unpredictable) developments. It permits ambiguity (variability) within a certain (regular) dynamic. Because of this ambiguity, a complex system can be as it is, but at the same time it can be completely different.[12]

Unlike the structural complexity of structures, in which the integration of various elements is treated as a pattern,[13] the concept of dynamic complexity relates to a functional whole within this context of the integration of various dynamics. A machine is not complex because it consists of various parts, but, at most, because it integrates the dynamics of these parts and unifies them in a function. It is precisely in this sense that modern products have become complex, as the example of fuel injection has shown.

However, modern technologies do not just have to integrate technical appliances into a function. They also have to meet economic, political and social demands. Overall, today's new products have to:

- *Be economically viable.* In other words, it has to be possible to develop and produce them with the available commercial resources. They have to be able to assert themselves on the market and they also have to make a profit.
- *Be technologically feasible.* This means they have to function in technological terms while simultaneously being safe and reliable.

- *Fit into the political landscape*. In other words, they must not contravene existing environmental and safety standards.
- *Be accepted by customers*. This means they must fit into a real application context.

Because these functions cannot be met independently of each other, innovations become increasingly more complex. Changes to comply with technological safety may pose a risk to economic viability, new environmental standards may endanger technological functionality and technological specifications may threaten customer acceptance. Moreover, because of the functional unity, the whole cannot be broken down into its component problems and partial solutions cannot be integrated into a new product. Exactly this is the reason why the concept of complexity has become a decisive characteristic of modern innovations which requires the integration of different forms of competence for its production. Self-organising innovation networks offer an adequate measure to reduce the complex dynamics of innovation processes.

Self-Organisation: The Emergence and Maintenance of Order

The question that emerges in relation to innovation networks is, how do heterogeneous (individual or collective) actors form a network with its own autonomous way of operating that distinguishes itself from the environment and is broadly resistant to forms of external control without the rules for this 'momentum' being given externally? A formally satisfying, model-like answer to this question of the emergence of global structures from local interactions became possible in the 1960s. Because this involved completely different disciplines such as physics, chemistry, biology, sociology, psychology and linguistics, several related terminologies emerged, whose precise relationship with each other cannot be dealt with here. These included autopoiesis, dissipative structures, synergetic, self-reference and self-organisation. In the following, we shall use the concept of self-organisation as a generic term for all models addressing the emergence of order and its maintenance when faced with random interferences.

The literature provides many examples of self-organisation in physical and chemical systems (see, for example, Haken 1978 and Küppers 1996) that all have one thing in common: the cyclic linkage of non-equilibriums with the compensation processes that cause them. Non-equilibriums are understood here in general terms as unequal distributions of mass and/or energy that strive toward an equalisation according to the second fundamental law of thermodynamics; in other words, that lead to material and/or energetic compensatory currents. The decisive aspect is that these compensatory currents modify their cause – the non-equilibrium.[14]

This 'circular causality' between non-equilibrium as cause and compensation process as effect is the organisational principle of self-organisation.[15] A certain form of non-equilibrium generates precisely that compensation process that reproduces this non-equilibrium. The order this gives rise to is a dynamic state. Its maintenance requires the retention of the non-equilibrium in order to prevent the transfer into an unstructured, static equilibrium. Self-organised order therefore requires an environment that provides the resources for maintaining such non-equilibriums. Thus, the self-organisation of physical or chemical processes has to be open for the input of energy or matter. In biological or social orders, this can also be called openness for information (see especially Eigen and Schuster 1977, as well as Krohn and Küppers 1989).

Because of the non-linear dynamics of such compensation processes, several forms of order (structures) are always possible, with the specific order structure that emerges being determined solely by the strength of the non-equilibrium. The latter is measured by means of a so-called control parameter.[16] However, it is not its continuous change that causes a change in the pattern. It is far more the case that there are critical values beyond which the pattern changes suddenly. The conditional relationship between cause and effect shapes itself anew.[17] New order states then appear suddenly when critical values in the control parameter are exceeded.

Characteristic for this transition from one order state to another is the phenomenon of bifurcation. At a certain critical value in the control parameter, one state splits into two possible states. These then differ according to, for example, their symmetry.[18] Because of the repeated bifurcations that occur when the control parameter is raised even higher, the sequence of order states becomes path-dependent. This means that each time the control parameter runs (in the same way) through its value domain, another sequence of structures may emerge. The experimental design does not provide the observer with any information about which path sequence of structures this will actually be. The structure becomes an 'historical object' because the structure formation process is dependent on its history (see Prigogine 1980).

The models of self-organisation found in physics are particularly suitable for clarifying the principles of self-organisation, that is, its preconditions and its mechanisms. They can be presented in mathematical terms and tested experimentally. Important generalisations can be derived from them for other phenomenal domains such as that of living things or the social world. Following the generalisations taken from the natural sciences, the 'mechanics' of self-organisation can be summed up in the following principles (for more details see Küppers 1998):

- In general, the driving force of self-organisation is a dynamic situation which is governed by 'circular causality'. A cause produces an effect which changes its cause. In natural systems such a circular causality emerges if a given 'non-equilibrium' drives its 'compensation' and the compensation changes its cause, that is, the non-equilibrium.
- The circular causality as the driving force of self-organisation needs external resources to avoid the compensation of the cause by its effect, that is, by the dynamics caused. In the case of a non-equilibrium the compensation that leads to a (static) equilibrium must be avoided by recourse to external resources, that is, by the input of energy and/or matter.
- If this circular linkage of 'cause' and 'effect' is limited in space, this closed dynamics produces within this area a differentiation of a system (internal) and its environment (external) which can be observed. In this sense the emergence of a system is the result of circular causality.
- In non-material systems self-organisation produces boundaries. A distinction is made between the specific (circular) network of operation of the system and non-specific operations within the environment.
- If a cause generates precisely the effect that reproduces its cause, the dynamic comes to a 'stop', and a state of order with some degree of complexity has been established.
- If the circular relationship between a cause and its effect is non-linear, cause and effect no longer have a unique relationship. The relationship becomes ambiguous. One type of cause finds several forms of compensating effects. In natural systems a specific 'non-equilibrium' may have different patterns of 'compensation'. Different types of order are possible and the predictive power of linear theories is lost.
- Therefore, the emergence of order is a selection process in which a certain form of dynamic suppresses all others because it is the only one in which its cause is reproduced by the dynamic caused by it. The selection mechanism of self-organisation is the circular conditional relationship of cause and effect.

A successful application of the concept of self-organisation to the phenomenal domain of the social assumes that one can find social mechanisms exhibiting the form of circular causality. A social cause has to produce social effects that change their cause. A stable social order can then be understood as a reproduction of a social cause through the social effects that it triggers. The (social) perception of uncertainty and the social activities undertaken to reduce it possess such a circular structure, and are therefore proposed as a mechanism of social self-organisation (Küppers 1999).

Circular causality is the mechanism of self-organisation and leads to the emergence of order. Although a specific order always emerges only in a certain domain, this domain has no borders. For example, the Gulf Stream is restricted to a certain part of the Atlantic Ocean, but it has no exactly defined border. Borders only become important as a demarcation in biological organisms. Cells, as their smallest units, have cell walls as borders and the organisms constructed from them have specially equipped groups of cells – the skin – that help them to separate themselves from their environment. These borders ensure the cohesion of the parts from which the organism is composed and define its identity. They have however to be open for the resources that are needed for the internal dynamic of the organism. Whereas the borders of biological organisms are, in turn, complex organisms themselves, formed from specialised elements of the organism, the border of a social organisation generally does not consist of certain members who have been assigned the task of demarcation from the environment. The demarcation of a social organisation and thus its border is determined solely by the participation of its members in the specific operations of a specific organisation.

Systems: The Emergence of Boundaries

Borders mark edges between an 'internal' process organisation and an 'external' environment. They are the outcome of a differentiation through which a system distinguishes itself from its environment. To facilitate understanding of this exclusion of an environment, the following section will sketch systems and their properties, particularly within the context of self-organisation.

Systems are dynamic units. They are 'put together' from elements that interact (dynamically) with each other. However, neither the properties of the elements nor their interaction determine the properties of the system. Systems possess properties that their elements and their dynamics do not possess. When trying to solve the riddle that guarantees the integration of the parts with the unity of the whole, all kinds of occult principles such as 'nisus formativus' (Blumenbach), 'vis essentialis' (C.F. Wolff), 'Entelechie' (Driesch), 'moule intern' (Boffon) 'elan vital' (Bergson) or 'morphogenetic fields' (Sheldrake) were used. A more satisfying solution to the question of the relationship between the whole and its parts was found with the development of concepts like self-organisation.

One of the first breakthroughs in formulating a theory of 'self-organising systems' came about only in the 1920s with the beginnings of a general systems theory. One of its founders was Ludwig von Bertalanffy. As a biologist, he was interested in a theory of living systems. 'Because the fundamental character of a living thing lies in its organisation, the usual examination of single parts and single processes can provide no complete explanation of the phenomenon of life' (von Bertalanffy 1972). The object of systems theory could no longer be the analysis of linear causal chains of cause and effect, but the complex (non-linear) interaction of elements. What proved to be decisive for the further development of systems theory was the introduction of the concept of the 'open system' as a precondition of an environment. This term replaced the traditional difference between the whole and the part with the difference between system and environment. Open systems exchange energy, matter and information with their environment. Closed systems, in contrast, are isolated from their environment and, in the ideal case, they have no relationships of exchange with it. They strive toward a (thermodynamic) state of equilibrium in which all existing orders and differentiations disappear. Such equilibriums cannot be modified by the environment either, because of their isolation.

Open systems, in contrast, because of the continuous input of resources, are in a permanent state of non-equilibrium in which stationary states are possible only as the 'dynamic equilibriums' that Bertalanffy called *Fließgleichgewichte* (stationary states). Stationary states are dynamic states that do not change over time and that – once attained and given their stability – the system is unable to leave again by itself. After a deflection caused by perturbation from the environment, they will return to this state. This return to the initial state is solely a consequence of the system operations and thereby an autonomous achievement of the system. Systems that are capable of attaining such states are also able to maintain themselves in a variable environment. Stationary states possess an attractor property that functions as a quasi-teleological force. Because of a stationary state's stability, all neighbouring states that lie within the 'area of attraction' move toward it. Bertalanffy uses the term 'equifinality' to describe this. Stationary states are attained 'without this requiring the intervention of a regulating entelechy or soul' (von Bertalanffy 1972). According to Bertalanffy, open systems are systems that are able to organise their self-maintenance in relative independence of changes in the environment.

This characterisation of open systems brings Bertalanffy close to the self-organised systems that Heinz von Foerster (1960) introduced at the beginning of the 1960s. In his paper 'On self-organizing systems and their environments',[19] he emphasised the need for an environment as a 'condition *sine qua non*'.

This term 'self-organisation' becomes meaningless, 'unless the system is in close contact with an environment, which possesses available energy and order [sic], and with which our system is in a state of perpetual interaction, such that it somehow manages to "live" on the expenses of the environment' (ibid., p. 36).[20]

The decisive aspect in his model in the sense of the cybernetic paradigm is the aspect of self-regulation, that is, the circular causal linkage between cause and effect. Because of this circularity of their dynamics, self-organising systems cannot be controlled directly by their environments. They do not react directly to changes in the environment, but always in line with their 'eigen-dynamic'. Although even in simple non-linear systems, a pattern formation can be provoked through impulses from the control parameters, this cannot determine the pattern itself.[21]

Unlike the control switch on a machine, control can be built up only through knowledge of the eigen-dynamic of the organism and the existence of a structural relation between the environmental stimulus and reaction generated through feedback. This has far-reaching consequences: in the case of psychic systems for instance, learning is possible only if the learner offers the teacher a selection of different patterns of behaviour based on his or her eigen-dynamic so that the teacher may choose one and stabilise it through reinforcement mechanisms.

This is why self-organising systems are also called operationally closed. The term emphasises that all causes of changes to the state of the system are exclusively consequences of system-internal operations. This operational closure does not contradict the openness to resources of dynamic systems. It is far more the case that the difference between open and closed exists on two different levels. Thermodynamically, the difference is that between isolation and exchange with the environment – in this sense, dynamic systems are open. Cybernetically, in contrast, the open/closed difference is used in another way, and systems are open if they do not have feedback, that is, if their output does not become the input of a control loop. Closed systems, in contrast, are feedback systems with closed loops. Negative feedback is self-regulating; positive feedback generates system changes. In linear cases, these are processes of growth and shrinkage processes; in non-linear cases, new system states come about through instabilities (reinforcements of deviation). Hence, self-organising systems are thermodynamically open and, cybernetically, closed (see Luhmann 1991). They are 'historical'. Their behaviour in the present is determined by the past.

The mechanism of operational closure is simultaneously the mechanism for excluding an environment. In both operational and spatial terms, it defines a border marking the separation between system and environment. Systems are now no longer entities that an observer with specific epistemological interests has isolated from a total context, but process networks that separate themselves from the environment. Self-organising systems are, in this sense, real systems.

The self-organisation of the system occurs within its borders. This guarantees the system's identity even when structures change. Although organisation (circular causality) and structure determine each other, the conditional relationship can take different patterns. When organisation – and thereby, identity – remains constant, systems can adapt to a changed environment through structural changes. Adaptation is an achievement of self-organisation and not the outcome of selection through an environment.

The Self-Organisation of the Social

A successful application of the concept of self-organisation to the phenomenal domain of the social assumes that one can find social mechanisms exhibiting circular causality. A social cause has to produce social effects that change their cause. A stable social order can then be understood as a reproduction of a social cause through the social effects that it triggers. The (social) perception of uncertainty and the social activities to reduce it possess such a circular structure, and are therefore proposed as a mechanism of social self-organisation (see Küppers 1999).

In general, deficits in the regulation of social interactions are potential sources of social uncertainties. These can emerge through, for example, the introduction of new technologies with new risks that can no longer be regulated with previously established social practices. This new technology reveals gaps in regulation or makes existing regulations seem inadequate. But it is not only a new technology which requires social regulation for its use. There is an unavoidable social risk within social interaction which, in general, needs regulation: the risk of violence within confrontation and the risk of being cheated within co-operation. The regulation of societal co-operation thus becomes a permanent feature of social coping with uncertainty, and society becomes a dynamic system.

This way of dealing with uncertainty is also the mechanism involved in the construction of the social as an autonomous phenomenal domain. It supposes a specific micro-sociological linkage mechanism: Everybody acts in anticipation of the acts of others while taking into account their own conjectures regarding what others expect of them. Hence, the decisive mechanism for the macro-sociological pattern formation is, once again, a feedback between uncertainty (cause) and the coping with this uncertainty (effect) on the micro-level that leads to the formation of social rules, and after a while, to a consolidation in the form of the institutionalisation of social interaction in terms of social institutions.

Through the mechanism of 'circular causality' the social dynamics becomes 'closed' because (social) causes and their (social) effects change each other. As a consequence a specific form of uncertainty and its regulation begins to reproduce each other. A specific social structure has emerged as a 'stationary state' of the social dynamics, that is, the perception and regulation of social uncertainties. The emerging social structures consist of action rules that consolidate conservatively into the various institutions of society such as the legal system, business organisations, networks or political parties. Although the general possibility of contingency, of being able to be other, is retained in principle, it becomes powerless when faced with the conservative power of this institutional arrangement toward which all action has always been oriented in a given society.

The manner of dealing with social uncertainty determines the forms of social self-organisation. Individual behaviour is integrated to form social patterns without any need for directives from the environment. Examples of this are the large functional sub-systems of society: politics, law, the economy, science and so forth. They have all emerged as rule systems of social action because they have proved their worth in functional terms: they have reduced the social uncertainty in their own domain through regulations laid down in the specific systems of institutions. For example, police, public prosecutors, judges and lawyers investigate, press charges, adjudicate and defend according to rules that are accepted within a society and that are considered appropriate to guarantee public safety and social justice.

But social uncertainties cannot be transformed into social certainty. Rules (consensus, agreements, contracts, laws, moral obligations) can only reduce them to a socially acceptable form of relative certainty for a limited period of time, which dynamic societies may once again use as the starting point for new forms of uncertainty. Therefore, social uncertainty exists, not as an objectively given state of affairs, but as something that is constructed through social negotiations. Experience, knowledge, cultural practices and the like thus take on an important role. This context opens up scope for interpretation within which the perception and definition of social uncertainty as well as procedures for dealing with it have to be negotiated socially.

Processes of social self-organisation deliver the selection mechanisms to which such negotiation processes are subjected. They replace the classical idea of an objectively given rationality inherent to the nature of things that permits an unequivocal definition of and an appropriate way of dealing with uncertainty. This socially constructed aspect of risk was long overlooked in the political discussion on technological risks in, for example, the controversy over nuclear energy. Nowadays, the 'nature of the thing' proves to be, itself, a social construction that can be maintained only as long as the belief in the 'mechanistic world view' seems to be justified through the successes of modern science.

Innovation Networks

As mentioned in the first section of Chapter 1 of this volume, innovations are becoming increasingly more complex. In most of the products an increasing number of parts is integrated into a single function. At the same time a product has to fit into the market, that is, to fulfil the specific requirements of a potential user. 'This complexity has meant that multidisciplinary knowledge has become necessary for the generation and development of new products. In the computer industry, for example, the disciplines involved in the innovation process may range from solid state physics to mathematics, and from language theory to management science' (Malerba 1992).

But it is not only this 'multi-' or 'trans-disciplinarity' which means that innovations are products that are burdened with uncertainty. Generally speaking, there is a lack of knowledge on how to transfer even a known technology into a new usage context. Although the underlying effects are known, nothing is known about how they will behave in the planned application context. One example is that even though a lot is known about the physics of fusion processes, the construction of a fusion reactor would be a lengthy and expensive enterprise. Indeed, throughout the world, research teams have been trying to produce the necessary knowledge for more than 40 years. Although this is an extreme case, it illustrates the knowledge uncertainties that may be linked to the use of still existing knowledge within the context of a new technology.

When the application context is new, there are no databanks, no experts whose knowledge can be drawn on. This knowledge first has to be developed. The starting point for the production of new knowledge are hypotheses formulated on the basis of available knowledge. These hypotheses prove to be either right or wrong when tested on a prototype, in a market analysis or a customer survey. If a hypothesis is not confirmed, it is modified and retested. This testing is itself a complex process, because it is not simply a case of comparing data with hypotheses. Instead – once again, with the aid of knowledge – a decision is made on how far the chosen procedures are appropriate for testing the hypotheses, and in what way the data have to be interpreted in order to confirm them (see Krohn and Küppers 1989 and Küppers 1998). In this sense, knowledge production is a series of hypotheses and the data they generate determine each other as new knowledge that is believed to be certain. In the simulation model this is implanted as an innovation oracle (see Chapter 4).

For this reason the construction of an innovation requires far more intellectual, social and material resources than are generally available to a single company (see, for example, the BioTech case study in Chapter 4 of this volume).

Companies therefore have to co-operate with other companies in order to gain access to the resources they lack. In many cases, they also have to co-operate with non-business institutions. These are mostly research centres at universities and government laboratories as well as the state-run and non-state-run institutions that regulate and license new technologies. This is because innovations must not only function technologically and be economically viable, but also fit into the socio-political environment. In the innovation networks built up over such co-operative relationships, resources necessary for the success of the innovation process can be deployed: theoretical and practical knowledge as a precondition for the development and implementation of innovations, social competence for the successful organisation of co-operation in the network, and the necessary financial means to pay for it all.

Therefore, in innovation networks discursive negotiations aimed at reducing the uncertainty of the innovation process with respect to its different dimensions serve as a co-ordinating mechanism. This differs from the formal contracts that dominate market co-ordination, as well as from the principle that instructions co-ordinate social interaction within hierarchical organisations (Kowol and Krohn 1995). Therefore, innovation networks differ from classical organisations in terms of their structure and dynamics. Although they can be set up as the outcome of contractual regulations between the co-operating institutions, their internal form of operation is neither regulated contractually nor determined by management directives. Innovation networks operate according to the principles of self-organisation. The element that triggers their emergence or establishment is the idea that the uncertainty making the development of innovations a risky venture as a result of increasing complexity can be reduced by co-operating with suitable partners. This trust in the reciprocal utility of co-operation is the integrating force within the network. As a rule, it is stronger than the fear of being disappointed and deceived by one's partners. Building up this trust and overcoming the fear are preconditions for the emergence of a network and determine its dynamics.

Innovation networks reduce different kinds of uncertainties in the innovation process with respect to its technological, social and economical dimensions. This calls for the integration of various competencies. Which competence is relevant is not certain at the outset. It is only during the course of the innovation process that the areas in which knowledge is uncertain become apparent. Therefore, as a rule, innovation processes are dynamic, in other words, they modify their co-operative relationships over the course of time. This does not mean that membership of a network cannot be regulated through contracts or job delegation. However, the network itself decides which members are involved in which forms of co-operation. Innovation networks are determined through their operations – the production of knowledge, the integration of diverging interests – and not through formal memberships. Although they may influence the operation as marginal conditions, they cannot finally determine it.[22]

Besides the problem of social integration, knowledge production in innovation networks reveals the characteristic features of self-organisation: circular causality, autonomy and self-reference. Knowledge uncertainty is reduced, in other words, modified through knowledge, and the modified knowledge uncertainties require new knowledge for their reduction. At the end, the network comes to the conclusion that it knows everything it needs to know, it possesses all the knowledge, and can implement it in a new product or process. As a result, innovation is generally not an artefact, but a recipe for its manufacture. It covers technological details, the conception of exploitation contexts and extrapolations relating to commercial success.

In this sense, the production of new technologies is no longer a process of trial and error involving variation and selection but, rather, a co-ordinated process of knowledge production and learning. This process involves actors such as suppliers, banks, universities, research institutes and governmental regulatory bodies. The number of actors involved is determined by several dimensions of uncertainty associated with innovations: time frames, technological complexity, modalities of financing, deficits in knowledge, legal issues, risks and public sensitivity (Braczyk 1997; Krohn 1997b).

To summarise, the formation of new innovation regimes will depend increasingly on whether, and to what extent, actors are able to construct flexible networks that exploit the advantages of co-operation and banish the disadvantages of rigid organisations and turbulent markets. This is thought to be the theoretical background for the simulation model presented in Chapter 7.

NOTES

1. For a comparison with the classic controversy between supply induction versus demand induction, see Kowol and Krohn (1994) and Kowol (1998).
2. For a discussion of the concept of evolution with respect to innovations see, for example, John Ziman (2000).
3. This also includes BETA (University of Strasbourg, France), IKE (University of Aalbourg, Denmark), INSIDE (University of La Sapienza, Rome, Italy), MERIT (University of Limburg, the Netherlands), and SPRU (University of Sussex, UK).
4. The dominant design concept was introduced in the pioneering work of Abernathy and Utterback (1975). Since then it has also been used in other scientific disciplines (sociology, population ecology and game theory).
5. See, for example, the various articles in the *Journal of Evolutionary Economics* (1999).
6. This concept can be attributed to Paul Windrum.
7. See the various articles in Rammert (1994) and Rammert (1997); see, for research on the production of new technologies in the nations of the European Union, Cronberg and Sørensen (1995).
8. These visions of use also contain the entrepreneur's expected profits.
9. Nonetheless, it has to be considered that these drafts may go wrong for a variety of reasons (higher production costs, unfavourable market constellations, etc.). Hence, all innovation processes have an anticipatory character.
10. Although the term innovation process implies a dynamic, it is necessary to identify the driving force, or otherwise the notion dynamics is meaningless.
11. Empirical examples of such learning processes can be found in Wildemann (1994), Probst and Büchel (1998), von Hippel and Tyre (1995) and Kowol (1998).
12. This example uses the title of a book by Helga Nowotny (1999): 'It is as it is, but it could always be different.' [Es ist so. Es könnte auch anders sein.]

13. Compare with terms such as entropy, redundancy, and the like.
14. In physics, there are two different possibilities of dynamics. In mechanics, a force ensures a movement. In thermodynamics, a disequilibrium strives toward compensation. Self-organisation is often viewed as linking the two principles. The movement is ordered, the compensation is diffuse. The order emerges through the interplay of both principles.
15. As already ascertained by Kant. See Krohn and Küppers (1992).
16. In the case of a convection current in which an upper/lower temperature difference in a fluid layer drives a compensation of the difference in temperature through a current, it is precisely this temperature difference that determines the transition between the different forms of current.
17. In the example of cellular convection, different flow patterns are found that become increasingly more 'complex' the greater the difference in temperature (control parameter). Typical flow patterns are 'rolls' (the fluid flows in the form of horizontal scrolls of a certain thickness with alternating direction of rotation), squares (the fluid moves from bottom to top in the centre and downward at rectangular edges with the direction of rotation of the movement alternating from square to square), or the more complex hexagonal flow.
18. In the case of scroll flow, the direction of rotation of the scroll is undetermined. At the critical points, chance in the form of small random interferences decides which of the two options emerges.
19. The 'order-from-noise' mechanism presented in this paper, which Prigogine later called 'order from fluctuation', gave rise to a number of fallacies. Naturally, in this context, we do not mean that disruptions or fluctuations form the 'raw material' from which the order is 'made'. Instead, we mean that disruptions can trigger instabilities; that is, destabilise a certain state so that non-order transforms to order or one order transforms into another. The mechanism for the emergence of order is circular causality.
20. The second condition formulated by Heinz von Foerster was that the environment has to be structured. See von Foerster (1984, p. 36).
21. A common misunderstanding is neglecting a difference between inducing and determining.
22. The situation is different when balancing out diverging interests. In this case, members are not free from the interests of the institutions they represent. Trust in the utility of the network not only has to exist in the network itself but must also be capable of being communicated to the outside. It is those outside, who ultimately provide the resources for the network, who have to be convinced at the end of the day.

REFERENCES

Abernathy, W. and M. Utterback (1975), 'A dynamic model of product and process innovation', *Omega*, **3**, 639–56.

Anderson, E.S. (1991), 'Techno-economic paradigm as typical interface between producers and users', *Journal of Evolutionary Economics*, **1**, 119–44.

Anderson, P. and M.L. Tushman (1990), 'Technological discontinuities and dominant design: A cyclical model of technological change', *Administrative Science Quarterly*, **35**, 604–33.

Argyris, C. and D.A. Schön (1978), *Organizational Learning: A Theory of Action Perspective*, Reading, MA: Addison-Wesley.

Aristotle (1959), *Die Politik*, Paderborn: Schönigh.

Arrow, K. (1962), 'The economic implications of learning by doing', *Review of Economic Studies*, **29**, 155–73.

Bertalanffy, L. von (1972), 'Vorläufer und Begründer des Systemtheory', in R. Kurzrock (ed.), *Systemtheory*, Berlin: Colloquium, pp. 17–27.

Bijker, W.E., T.P. Hughes and T.J. Pinch (eds) (1987), *The Social Construction of Technological Systems. New Directions in the Sociology and History of Technology*, Cambridge, MA: MIT Press.

Birchenhall, C., N. Kastrinos and M. Metcalfe (1997), 'Genetic algorithms in evolutionary modelling', *Journal of Evolutionary Economics,* **7**, 375–91.

Braczyk, H.J. (ed.) (1997), *Innovationsstrategien im deutschen Maschinenbau – Bestandsaufnahme und neue Herausforderungen*, Arbeitsbericht, Stuttgart: Akademie für Technikfolgenabschätzung in Baden-Württemberg, p. 83.

Brenner, T. (1998), 'Can evolutionary alogrithms describe learning process?', *Journal of Evolutionary Economics,* **8** (XXX), 271–83.

Callon, M. (1992), 'The dynamics of techno-economic networks', in R. Coombs, P.P. Saviotti and V. Walsh (eds), *Technological Change and Company Strategy. Economic and Sociological Perspectives*, London: Academic Press, pp. 72–102.

Callon, M. and J. Law (1989), 'On the construction of sociotechnical networks: Content and context revisited', *Knowledge and Society: Studies in the Sociology of Science Past and Present*, **8**, 57–83.

Campbell, D. (1965), 'Variation and selective retention in socio cultural evolution', *General Systems,* Yearbook **14**, 69–85

Cronberg, T. and K.H. Sørensen (1995), 'Similar concerns, different styles? Technology studies in Western Europe', *Proceedings of the COST A4 workshop in Ruvaslathi, Finland, European Commission, Social Science*, **4**, Brussels and Luxembourg.

Daele, W. van den, and W. Krohn (1998), 'Experimental implementation as a linking mechanism in the process of innovation', *Research Policy*, **27**, 853–63.

David, H. and M. Kopel (1998), 'On economic application of the genetic algorithm: a model of the cobweb type', *Journal of Evolutionary Economics*, **8** (3), 297–315.

Dosi, G. (1983), 'Technological paradigms and technological trajectories. The determinants and directions of technical change and the transformation of the economy', in C. Freeman (ed.), *Long Waves in the World Economy*, London: Butterworth, pp. 78–101.

Dosi, G. and J.S. Metcalfe (1991), 'Approches de l'irréversibilité en théorie économique', in R. Boyer (ed.*), Les figures de l'irréversibilité en économie*, Paris: Editions EHESS, pp. 37–68.

Dosi, G. and R.R. Nelson (1994), 'An introduction to evolutionary theories in economics', *Journal of Evolutionary Economics*, **4**, 153–72.

Eigen, M. and P. Schuster (1977), 'The hypercycle – A principle of natural self-organization', *Die Naturwissenschaften,* **64**, 541–65.

Foerster, H. von (1960), 'On self-organizing systems and their environment', in M.C. Yovits and S. Cameron (eds), *Self-Organizing Systems,* London: Pergamon Press, pp. 31–50.

Foerster, H. von (1984), *Observing Systems*, Seaside, CA: Seaside.

Foster, J. (2000), 'Competitive selection, self-organisation and J.A. Schumpeter', *Journal of Evolutionary Economics,* **10** (3), 311–28.

Freeman, C. (1991), 'Network of innovators: A synthesis of research issues', *Research Policy*, **20**, 499–514.

Goldberg, D.E. (1989), *Genetic Algorithms in Search, Optimization, and Machine Learning*, Reading, MA: Addison-Wesley.

Grabher, G. (ed.) (1993), *The Embedded Firm. On the Socioeconomics of Industrial Networks*, London: Routledge.

Haken, H. (ed.) (1978*), Synergetics: An Introduction. Nonequilibrium Phase Transitions in Physics, Chemistry and Biology, 2*, Berlin: Springer.

Halfmann, J. (1995), 'Kausale Simplifikation. Grundlagenprobleme einer Theorie der Technik', in J. Halfmann, W. Rammert and G. Bechmann (eds), *Jahrbuch Technik und Gesellschaft*, 8, Frankfurt/M. and New York: Campus, pp. 211–26.

Herbold, R., W. Krohn and M. Timmermeister (2000), 'Innovationsnetzwerke – Organisationsbedingungen für Innovationsdynamik und Demokratie', in R. Martinsen and G. Simonis (eds), *Demokratie und Technik. (k)eine Wahlverwandschaft?*, Opladen: Leske und Budrich, pp. 225–47.

Hippel, E. von, and M.J. Tyre (1995), 'How learning by doing is done: Problem identification in novel process equipment', *Research Policy*, **24** (1), 1–12.

Holland, J.H. ([1975] 1992), *Adaptation in Natural and Artificial Systems*, A Bradford Book, Boston, MA, and Cambridge, MA: MIT Press.

Hughes, T.P. (1983*), Networks of Power. Electrification in Western Society 1880–1930*, Baltimore: John Hopkins University Press.

Hughes, T.P. and R. Mayntz (eds) (1988), *The Development of Large Technical Systems*. Frankfurt/ M.: Campus.

Johnson, B. (1992), 'Institutional learning', in B.-A. Lundvall (ed.), *National System of Innovation. Toward a Theory of Innovation and Interactive Learning*, London and New York: Pinter, pp. 23–44.

Kämper, E. (1995), 'Die Organisation des technischen Wandels', *Der Beitrag der Organisationstheorie zur Innovationsforschung*, unveröffentlichte Diplomarbeit, Bielefeld.

Konrad, W. and G. Paul (1999), *Innovation in der Softwareindustrie. Organisation und Entwicklungsarbeit*, Frankfurt/M. and New York: Campus.

Kowol, U. (1998), *Innovationsnetzwerke. Technikentwicklung zwischen Nutzungsvisionen und Verwendungspraxis*, Wiesbaden: Deutscher Universitätsverlag.

Kowol, U. and W. Krohn (1994), *Innovationsnetzwerke*, Bielefeld: IWT-Papers # 6.

Kowol, U. and W. Krohn (1995), 'Innovationsnetzwerke. Ein Modell der Technikgenese', in W. Rammert, G. Bechmann and J. Halfmann (eds), *Jahrbuch Technik und Gesellschaft 8*, Frankfurt/M. and New York: Campus, pp. 77–105.

Kowol, U. and W. Krohn (1997), 'Modernisierungsdynamik und Innovationslethargie. Auswege aus der Modernisierungsklemme', in B. Blättel-Mink and O. Renn (eds), *Zwischen Akteur und System. Die Organisierung von Innovation*, Opladen: Westdeutscher Verlag, pp. 43–74.

Kowol, U. and W. Krohn (2000), 'Innovation und Vernetzung. Die Konzeption der Innovationsnetzwerke', in J. Weyer (ed.), *Soziale Netzwerke. Konzepte und Methoden der sozialwissenschaftlichen Netzwerkforschung. Lehr- und Handbücher der Soziologie*, Munich and Vienna: Oldenbourg, pp. 135–60.

Krohn, W. (1997a), 'Rekursive Lernprozesse, Experimentelle Praktiken in der Gesellschaft – Das Beispiel der Abfallwirtschaft', in W. Rammert and G. Bechmann (eds), *Technik und Gesellschaft, Jahrbuch 9: Innovation – Prozesse, Produkte, Politik*, Frankfurt/M. and New York: Campus, pp. 65–89.

Krohn, W. (1997b), 'Die Innovationschancen partizipatorischer Technikgestaltung und diskursiver Konfliktregulierung', in F. Gloede, L. Hennen and S. Köberle (eds), *Diskursive Verständigung? Mediation und Partizipation in Technikkontroversen*, Baden-Baden: Nomos, pp. 226–46.

Krohn, W. and G. Küppers (1989), *Die Selbstorganisation der Wissenschaft*, Frankfurt/M.: Suhrkamp.

Krohn, W. and G. Küppers (1992), 'Die natürlichen Ursachen der Zwecke – Kants Ansätze zu einer Theorie der Selbstorganisation', in W. Krohn, H.-H. Krug and G. Küppers (eds), *Konzepte von Chaos und Selbstorganisation in der Geschichte der Wissenschaften,* Jahrbuch Selbstorganisation Band 3, Berlin: Duncker & Humblot, pp. 31–50; and in G. Rusch and S.J. Schmidt (eds) (1992), *Konstruktivismus: Geschichte und Anwendung*, DELFIN, Frankfurt/M.: Suhrkamp, pp. 34–58.

Küppers, G. (ed.) (1996), *Chaos und Ordnung: Formen der Selbstorganisation in Natur und Gesellschaft,* Stuttgart: Reclam.

Küppers, G. (1998), 'The selforganization of social systems – A simulation of the social construction of knowledge', in P. Ahrweiler and N. Gilbert (eds), *Computer Simulations in Science and Technology Studies*, London: Springer, pp. 145–55.

Küppers, G. (1999), 'Coping with uncertainty – New forms of knowledge production', in K.S. Gill (ed.), *AI & Society*, London: Springer-Verlag, pp. 52–62.

Luhmann, N. (1991), 'Selbstorganisation und Information im politischen System', in U. Niedersen (ed.), *Jahrbuch für Komplexität in den Natur-, Sozial- und Geisteswissenschaften,* Berlin: Dunker & Humblot, p. 28 (translated).

Lundvall, B.A. (ed.) (1992), *National Systems of Innovation: Towards a Theory of Innovation and Interactive Learning*, London: Pinter.

McKelvey, M.D. (1996), *Evolutionary Innovations: The Business of Biotechnology*, Oxford: Oxford University Press.

Malerba, F. (1992) 'Learning by firms and incremental technical change', *Economic Journal,* **102**, 845–9.

Marengo, L. and M. Willinger (1997), 'Alternative methodologies for modelling evolutionary dynamics. Introduction', *Journal of Evolutionary Economics,* **7** (4), 331–8.

Mayntz, R. (1992), 'Modernisierung und die Logik von interorganisatorischen Netzwerken', *Journal für Sozialforschung*, **32** (1), 19–32.

Molina, A. (1999), 'The role of the technical', *Innovation and Technology Development: The Perspective of Sociotechnical Constituencies, Technovation, 1998*, **19**, 1–26.

Nelson, R.R. (1995), 'Recent evolutionary theorizing about economic change', *Journal of Economic Literature*, **33**, 48–90.

Nelson, R.R. and S. Winter (1982), *An Evolutionary Theory of Economic Change,* Cambridge, MA: Harvard University Press.

Nowotny, H. (1999), Es ist so. Es könnte auch anders sein, Frankfurt/M.: Suhrkamp.

Pohlmann, M. (1996), 'Antagonistische Kooperation und distributive Macht: Anmerkungen zur Produktion in Netzwerken', *Soziale Welt*, **27** (1): 44–67.

Powell, W.W. (1990), 'Neither market nor hierarchy: Network forms of organization', *Research in Organizational Behavior*, **12**, 295–336.

Powell, W.W. and P.J. DiMaggio (eds) (1991), *The New Institutionalism in Organizational Analysis*, Chicago and London: The University of Chicago Press.

Prigogine, I. (1980), *From Being to Becoming*, New York: Freeman.

Probst, G.J. and B.S. Büchel (1998), *Organisationales Lernen, Wettbewerbsvorteil der Zukunft*, Wiesbaden: Gabler.

Rammert, W. (1993), *Technik aus soziologischer Perspektive: Forschungsstand, Theorieansätze, Fallbeispiele. Ein Überblick*, Opladen: Westdeutscher Verlag.

Rammert, W. (1994), 'Die Technik in der Gesellschaft, Forschungsfelder und theoretische Leitdifferenzen im Deutschland der 90er Jahre', *Verbund Sozialwissenschaftliche Technikforschung*, **13**, 4–58.

Rammert, W. (1997), 'Auf dem Weg zu einer postschumpeterianischen Innovationsweise', in D. Bieber (ed.), *Technikentwicklung und Industriearbeit, Industrielle Produktionstechnik zwischen Eigendynamik und Nutzerinteressen*, ISF–München, Frankfurt/M. and New York: Campus, pp. 45–71.

Rammert, W. and G. Bechmann (eds) (1994), *Technik und Gesellschaft, Jahrbuch 7, Konstruktion und Evolution von Technik*, Frankfurt/M. and New York: Campus.

Rölle, D. and B. Blättel-Mink (1998), 'Netzwerke in der Organisationssoziologie – Neuer Schlauch für alten Wein?', *Österreichische Zeitschrift für Soziologie*, **23** (3), 66–87.

Rosenberg, N. (1982), *Inside the Black Box: Technology and Economics*, Cambridge: Cambridge University Press.

Rycroft, R.W. and D.E. Kash (1999), *The Complexity Challenge. Technological Innovation for the 21st Century*, London and New York: Pinter.

Scott, R.W. and J.W. Meyer (1991), 'The organization of societal sectors: Propositions and early evidence', in W.W. Powell and P.J. DiMaggio (eds), *The New Institutionalism in Organizational Analysis*, Chicago: University of Chicago Press, pp. 108–40.

Senge, P.M. (1990), *The Fifth Discipline: The Art and Practice of the Learning Organization*, New York: Klett-Cotta (deutsche Ausgabe: Stuttgart 1996).

Silverberg, G. (1988), 'Modelling economic dynamics and technical change: Mathematical approaches to selforganisation and evolution', in G. Dosi, C. Freeman and R. Nelson (eds), *Technical Change and Economic Change Theory*, London and New York: Pinter, pp. 531–59.

Summerton, J. (ed.) (1994), *Changing Large Technical Systems*, Boulder, COL: Westview Press.

Sydow, J. (1992), *Strategische Netzwerke – Evolution und Organisation*, Wiesbaden: Gabler.

Sydow, J. and A. Windeler (1997), 'Komplexität und Reflexivität – Momente des Managements interorganisationaler Netzwerke', in H.W. Ahlemeyer and R. Königswieser (eds), *Einfach komplex. Strategien, Konzepte und Beispiele*, Wiesbaden: Gabler, pp. 147–62.

Teubner, G. (1992), 'Die vielköpfige Hydra: Netzwerke als kollektive Akteure höherer Ordnung', in W. Krohn and G. Küppers (eds), *Emergenz: Die Entstehung von Ordnung, Organisation und Bedeutung*, Frankfurt/M.: Suhrkamp, pp. 189–216.

Tushman, M. and L. Rosenkopf (1992), 'Organizational determinants of technological change: Toward a sociology of technological evoiution', *Research in Organizational Behavior*, **14**, 311–47.

White, H.C. (1981), 'Where do markets come from?', *American Journal of Sociology*, **87**, 517–47.

Wiesenthal, H. (1995), 'Konventionelles und unkonventionelles Lernen: Literaturreport und Ergänzungsvorschlag', *Zeitschrift für Soziologie*, **24** (2), 137–55.
Wildemann, H. (1994), 'Das lernende Unternehmen. Weniger Verschwendung und Fehlleistung durch Lernen in der Organisation', *Technische Rundschau*, **29/30**, 14–18.
Williamson, O.E. (1990), *Die ökonomischen Institutionen des Kapitalismus*, Tübingen: Mohr.
Ziman, John (ed.) (2000), *Technological Innovation as an Evolutionary Process*, Cambridge: Cambridge University Press.

PART TWO

Case Studies

3. Innovation Networks by Design: The Case of Mobile VCE

Janet Vaux and Nigel Gilbert

This chapter records a moment in UK research funding practice involving the deliberate implementation of a novel form of innovation network. The networks in question are collaborative virtual research centres, that is, geographically distributed research centres involving both academic and industrial actors. As political constructs, they address issues of research excellence and of wealth creating R&D – two potent concerns in contemporary science policy discourse. The idea of Virtual Centres of Excellence (VCEs) was developed within the UK Foresight initiative, specifically in discussion in the Foresight Communications Panel, and was described in the panel report (Office of Science and Technology (OST) 1995b, para. 4.29). Two VCEs were set up as an almost direct result of the panel report – Mobile VCE in the field of mobile communications and Digital VCE in the area of multi-media and digital signal processing. Mobile VCE, at least, has been successful in terms of attracting industrial members and winning further research funding. However, these VCEs were the first and apparently the last of their kind. In this study of Mobile VCE we approach the larger questions of the value of this form of network, through the exploration of different perspectives on the purpose, originality and success of the network, based on interviews (carried out between March and November 1999) with a variety of individual actors associated with Mobile VCE.[1]

If Mobile VCE is an innovation network, to what extent may it be considered a self-organising network? Depending on one's perspective, the political actors that we describe in this case study may be viewed either as instigators without whom Mobile VCE would never have come into existence, or as providing environmental resources which the network successfully exploited. Mobile VCE – or to give its full name, the Virtual Centre of Excellence in Mobile Communications – was set up with the help of UK government funds under the Foresight Challenge Awards. However, government assistance in the formation of Mobile VCE was not confined to the provision of funds, since civil servants within the Department of Trade and Industry (DTI) actively facilitated the design of the research proposal. Mobile VCE was formally inaugurated in 1996, bringing together seven universities and more than 20 subscribing companies, including almost all the major players in the mobile phone industry in Europe. At this stage it became a self-governing organisation administered by a private company. At the completion of its initial three-year research programme, early in 2000, Mobile VCE succeeded in winning funds for a further three-year period from the Engineering and Physical Sciences Research Council (EPSRC), with a new research programme and a slightly enlarged academic membership, and continuing support from 20-plus industrial companies.[2] Its continuing ability to attract and retain corporate sponsorship is among the symptoms of its success. This study, then, is structured to take account of both the political and industrial contexts of the career of Mobile VCE.

THE POLICY CONTEXT

We suggested above that Mobile VCE may be described as a product or outcome of technology policy in the sense that it was developed within the UK Foresight initiative in the first half of the 1990s. However, the Foresight initiative was itself located within a tradition of historical analysis of recent technology policy, which may be seen both in academic commentaries and in policy discussion documents.

Within the academic literature on the subject, most authors agree that the conditions of production of science and technology are changing, with many closer links between academic and industrial research, and that the aims and scope of policy have altered to facilitate the economic and social exploitation of science and technology (for example: Gibbons et al. 1994; Faulkner and Senker 1995; Leydesdorff and Etzkowitz 1996). Despite some differences, the academic texts broadly agree on a history of changing concerns in technology policy, including an increased value placed on the exploitation of knowledge (based on assumptions about the power of technological innovation to effect economic and social change) and the encouragement of new forms of linking and networking to support increased technological innovation. It seems legitimate to speak of a 'discourse' of technology policy, in the sense of a set of terms and issues recognised and reproduced by 'insiders'. One leading scientist interviewed in 1998 (cited in Henkel et al. 2000) traced the changes in UK science policy from the early 1980s when 'researchers were encouraged to go and do something useful' to 'a sea-chang' in the late 1980s when the Thatcher Government decided that 'the proper place for government funding . . . was where there was market failure', that is, in basic science, and finally:

> In the last eight or nine years there has been the recognition that we had to become useful not by doing applied things but by networking and making our basic science available and exploitable. . . . [This] means that if you do basic science, maybe in strategic areas, you should make sure that anything that arises from that is exploited and the Intellectual Property Right is fed back.

One of the most significant recent developments in the implementation of UK science policy was the Foresight initiative. Within the policy literature, and especially in texts written by or for a policy-making audience, the term 'foresight' as an aim of policy became increasingly common during the 1980s and (in the UK) the 1990s. The issues were articulated in two publications by Ben Martin and John Irvine, the first commissioned by the Advisory Council on Applied Research and Development (ACARD 1986)[3], the second by the Dutch government (Martin and Irvine 1989). Martin and Irvine represented UK policy in comparison with other countries as failing to engage in a systematic 'forward look' at technological innovations. The idea of 'foresight' was distinguished from earlier ideas of picking strategic technologies,[4] which was recognised as too determinist and too linear, and was intended to incorporate perceptions of both technological and social trends, as a guide for policy makers in setting priorities.

The UK Foresight initiative was announced in the 1993 White Paper *Realising Our Potential* (ROP), the first major science policy document in the UK for 20 years (since the *Rothschild Report* in 1971). This document brought together science and industry policy, and echoes of ROP may be found in a subsequent series of industrial policy documents, the *Competitiveness* white papers.[5] The convergence between these policy areas is apparent in the opening paragraph of ROP (OST 1993, para 1.1):

> The understanding and application of science are fundamental to the fortunes of modern nations. Science, technology and engineering are intimately linked with progress across the whole range of human endeavour: educational, intellectual, medical, environmental, social, economic and cultural.

In its arguments for Technology Foresight, ROP mentioned both a need to properly utilise the UK's strength in science and technology and a desire to 'get maximum value for money' from public expenditure on science and technology.[6] In its announcement of Foresight, the White Paper stated:

> Technology foresight, jointly conducted by industry and the science and engineering communities, will be used to inform Government's decisions and priorities. The process will be carefully designed to tap into the expertise of people closest to

> emerging scientific, technological and market developments. The aim is to achieve a key cultural change: better communication, interaction and mutual understanding between the scientific community, industry and Government Departments.

The Foresight programme was structured to bring these groups together in its own committees and panels. It was implemented initially, in 1993, by the setting up of a steering group of industrialists, academics and government officials, and the steering group in turn instigated 15 sector panels, which again brought together industrialists, academics and government officials. Their remit was to consider the isses raised by social and other trends, and how they might be addressed by policy initiatives (see OST 1995a, para. 2.2). The sectoral panels set up by Foresight were expected to consider the specific conditions of their own areas, to identify strategic research and to make recommendations about how their sectors 'might best compete in the harsh global environment of the twenty-first century' (ibid., para. 3.2).

Mobile VCE as an Outcome of Policy

The Foresight Communications Panel approached the task of identifying global strategies for its sector with the experience of an important precedent: a LINK project called the Radio Frequency Engineers Education Initiative had brought together a consortium of industrial members and researchers at three universities – Bradford, Bristol and Surrey – to address gaps in radio engineering skills which were once again in demand for mobile communications. Individuals associated with the LINK project were also members of the Communications Panel. However, the collaborative mechanism devised by the panel, which they named Virtual Centres of Excellence (VCEs), were unlike LINK projects in several respects – in particular, they were to be focused on specific though 'virtual' research teams (LINK project funding is open to any academic-industrial collaborative team). The Panel recommendation stressed that VCEs would produce globally competitive research teams from existing 'small dedicated research teams in particular niche technologies' (OST 1995b, para. 4.29):

> [T]he scale of effort and rate of progress needed to compete internationally in mainstream technologies necessitates that we develop ways of harnessing the academic research effort of distributed teams and, where possible, linking these to industrial enterprises both large and small.

The Panel used the term 'virtual' to mean 'geographically distributed', that is, bringing together existing small teams. The rationale for this element of the recommendation, the need for world-class research teams, addressed a perceived problem of the fragmentation of research in the UK (OST 1995b, paras 3.34–3.36). The suggested solution, a virtual centre, was a way of bringing the fragments together without actually moving researchers from their institutions. This, it was felt, would avoid the financial and other costs of a bricks and mortar institution. As a representative of the DTI put it during an interview with us:

> Our experience of bricks and mortar institutions are not all that good. What tends to happen is you invest a lot of money in setting up a new building and within a short time the academics who really are the centre of expertise move on elsewhere, and you are left with maintaining an institution which is second rate.

In addition to providing a lower cost route to the construction of world-class research teams, this structure had implications for institutional collaboration. From the academics' point of view, it involved building a research team from institutions that had been used to competing with each other for research council and other funding, as well as in the various modern mechanisms of accountability, such as the Research Assessment Exercise. Similarly, for the companies, involvement in Mobile VCE implied some collaboration with competitors. This was an important point of difference between the VCE and the LINK programme approach which had previously been used as a funding mechanism in the sector. In a LINK programme, one or two universities and companies collaborate in a funding proposal; this was more open than Mobile VCE (as one interviewee put it, 'anyone could link up with anyone'), but also involved alliances of many fewer collaborators.

Among other recommendations, the Panel suggested that each team should have a fast wide area network connection (a recommendation that was overtaken by developments in internet use); that the VCEs should be interdisciplinary; and that each VCE should be run by a steering committee involving major collaborating companies, with a full-time co-ordinator. Finally, in accordance with its remit to identify key strategic areas, the Panel suggested four key areas that might be appropriate for setting up a VCE:

- Telecommunications software
- Broadband network architecture and design
- Mobile and personal communications systems
- Multi-media and digital signal processing.

Its mention in the Foresight Communications Panel recommendations was a crucial step in the history of Mobile VCE. It was intended (Parliamentary Office of Science and Technology (POST) 1994, para. 4.2) that each panel would develop a 'prioritised list' of technologies, and that the steering committee would then identify 'generic' technologies across the panels. However, the prioritised lists made by each panel also became important, not simply as the source of 'generic' technologies, but as a basis for implementing Foresight, which was a task for Foresight Phase II.[7] When the Foresight Challenge award was announced, bids were invited based on the recommendations of either the panel or steering group reports.[8]

Facilitating the setting up of VCEs became one way of implementing Foresight for those with an institutional responsibility for carrying out technology policy. Indeed the DTI began to explore the possibility of a VCE in the field of mobile communications as soon as the Foresight report (OST 1995a) published its list of priorities. A series of meetings was organised for the DTI by Professor John Gardiner of the University of Bradford. The initial meetings, to which industrial companies were invited, specified a number of target areas of research. Universities were then invited to a separate meeting and, Gardiner recalls, 'Surprisingly, or perhaps not so surprisingly, the same sort of consensus about research areas emerged among the university types.'

A funding proposal signed by Gardiner and Professors Barry Evans of Surrey and Joe McGeehan of Bristol (DTI 1996b)[9] eventually went forward as a successful bid to the Foresight Challenge awards (winning £1.5 million). This specified the four research areas that remained the 'core areas' of the first three-year research programme of Mobile VCE: Networks; Services and Service Metrics; Radio Environment; and Terminals.

In 1996 Mobile VCE was incorporated as a private company limited by guarantee (having no share capital)[10] to provide a legal entity for the management of the project and the holding of Intellectual Property Rights (IPR). It was funded collaboratively by industrial subscription and public (Foresight Challenge) money at a ratio of about 60:40. An Executive Director was hired in November 1996 to manage Mobile VCE in collaboration with an Executive Committee made up of representatives of the industrial and academic membership. If there is a point at which Mobile VCE became self-organising, this might be it. The political actors dropped into the background: both DTI and EPSRC had associate membership of Mobile VCE and sent observers to meetings, but project management was formally in the hands of the academic and industrial members.

This, however, is perhaps too literal a way of approaching the question of self-organisation. There are a number of slightly different respects in which Mobile VCE was an outcome of policy. On the one hand, the values and aims of policy discourse enabled the VCE to be produced as a solution. In addition, the work of policy-making institutions is apparent – both the work of temporary groups such as the Foresight Communications Panel, and the activity of civil servants charged with implementing policy in general and Foresight in particular. In particular, the meetings set up by the DTI were so important in bringing together the relevant industrial and academic representatives that more than one interviewee commented that 'Without the DTI it would not have happened.' This itself indicates the perceived political significance of VCEs at the time that Mobile VCE was being instigated.

MOBILE VCE AS A SELF-ORGANISING NETWORK

The regulatory framework and management structures set up for Mobile VCE provide a context in which the various actors associated with the organisation can relate to each other through both formal and informal links, facilitating 'networking' in more than one sense. The actors in Mobile VCE are primarily its academic and corporate members (through their various individual representatives) plus the executive director and government observers. Relations between the actors are partly mediated by the VCE's formal management structures, particularly through meetings of the steering committees of each of the core programme areas; but there are also a number of other meetings (Executive Committee, and plenaries on research results), as well as informal links between individuals. In this section we explore some of the ways in which relations between the various types of actors are negotiated through the formal selection mechanisms and management structures of Mobile VCE.

Selection Mechanisms

Mobile VCE has two main types of selection mechanism whereby the industrial and academic actors respectively gain membership of the VCE. Membership is open to companies on payment of a membership subscription available in 'shares' initially of £25,000[11], and to universities through a process of competitive selection. That is, formally speaking, the industrial members are self-selecting, while the number of academic members is limited and entry is controlled by the VCE's management structure. However, the processes whereby the founding members were recruited for the first core programme in 1996 show that some less formal selection mechanisms were also used.

Three universities (Bradford, Bristol and Surrey) were involved in the initial bid for Foresight Challenge Funding and were in effect selected by the OST's Challenge Award review process when it awarded a grant to Mobile VCE. At the time of the bid, it was proposed that a further two academic institutions might be selected to participate and in fact two new full members (King's College London and Strathclyde University) and two associate members (Edinburgh and Southampton) were then appointed by a VCE selection board from a field of 30 or 40 proposals.

So while a formal selection process was in operation, the VCE was also drawing upon a pre-existing network of representatives of industry and academia. Indeed, the initial prospectus pointed out that the three founding universities had previously been successful in a competitive bid for the DTI's 'Shortage of Radio Frequency Engineers' funds, that this research partnership had proved successful in the view of the industrial consortium involved, and:

> Since one of the purposes of the proposed Centre is to provide a collective means for industry to address advanced skills needs it would make sense to build upon this existing basis of co-operation. The DTI has taken soundings with industry and there was overwhelming support for proceeding in this manner. It has the incidental benefit of providing three universities to lead the bid under the Foresight Challenge (DTI 1996b, p. 10).

There is one other more general point of interest concerning the constraints on academic membership. One interviewee in our study contrasted the VCE selection process with the situation in the LINK programmes in mobile communications in which 'anyone could form a relationship with a company, or companies, or with other universities, and bid for money'. The result, in the case of Mobile VCE, might be described as a select club of those UK academic institutions recognised as being at the forefront of research in the field – that is, as a centre of *excellence*. There was little opportunity for recruiting further academic members, except at the commencement of a new programme, that is, a new three-year plan. Two new academic teams were indeed recruited in 2000 for renewed funding – Royal Holloway and a team originally from Queen Mary and Westfield College (QMWC) which was in the process of migrating to Southampton – who brought expertise in communications system security and software agents.

Centralised Management in Mobile VCE

As a collaboratively funded research project, one of the points of novelty about Mobile VCE was its incorporation as a private company and the appointment of a full-time director to co-ordinate the research programme. Mobile VCE is managed by an executive committee which includes the director and industrial and academic representatives. This committee makes all the important executive decisions: it has responsibility for the project management of the research programme as well as being the grant holder, drawing up the successful bid to the EPSRC for Core Two, the follow-on programme, which began in 2000. The administration of Mobile VCE by a company and its management by an executive committee have implications for both institutionalised and informal practices of accountability in the VCE. In particular, it is significant that the researchers report horizontally to consortium management, rather than vertically to outside funding bodies or companies. However, the executive committee has delegated many of its project management responsibilities to the steering committees of each of the core programme research areas. Also, although there are centralised mechanisms for the distribution of information, through technical meetings open to all the membership and the distribution of CD–ROM disks, the steering committee meetings provide important sites for communication between the academic and industrial members. The structure of the core programme and the workings of its component area steering committees therefore deserve some detailed description.

The Core Programme as a Management Structure

Research for the first three-year programme of Mobile VCE was organised around four core areas (Networks, Services, Radio Environment and Terminals) and the second is organised around three core areas (Software Based Systems, Networks and Services, and Wireless Access). Most of the research for this chapter was undertaken in the final year of the first three-year programme, and our description focuses on that. The four core research areas of the first programme were established early enough to be incorporated in the bid to Foresight Challenge and took account of the areas of research strength of the founding academic members. Research in each of the core areas was led by an academic co-ordinator and managed by an industrially-led steering committee, and three or four other university teams were also involved in each area, as below:

- *Networks*
  Academic co-ordinator: University of Surrey
  Other members: Bradford, Bristol, Strathclyde, King's College London
- *Services and Metrics*
  Academic co-ordinator: University of Bristol
  Other members: Bradford, Strathclyde, Surrey
- *Radio Environment*
  Academic co-ordinator: University of Surrey
  Other members: Bradford, Bristol, King's College London
- *Terminals*
  Academic co-ordinator: University of Bradford
  Other members: Bristol, Surrey, Edinburgh and Southampton.

The research areas each included a majority of the participating academic institutions, and each work area was managed collaboratively through quarterly steering group meetings led by industrial members. Thus the research areas acted as a mechanism to bring together academics from different institutions and representatives of the industrial members. The steering group meetings were the occasion for researchers to present ongoing results, and for industrial and academic members together to review progress and agree any changes needed in the direction of research. The steering committees were also responsible for overseeing intellectual property management, enforcing deadlines and monitoring progress.[12] However, it is worth noting that the relationship between industry and academic researchers was one of client/contractor. That is, industrial members were involved in processes of specifying, monitoring and approving research which was carried out by teams of academic researchers. This contrasts with LINK programmes, for example, which involve industrial and academic researchers working collaboratively.[13]

Virtuality of the Research Centre

The division of research among – rather than between – academic institutions was intended to encourage inter-university collaboration and, at the very least, all the academic collaborators met and communicated at their respective steering group meetings. As one of the founding academic members described the rationale:

> We wanted to make the virtual centre work as a centre. So whereas in a European project the programme organisers would agree what work packages were going to be done in what university and off you'd go and do your work packages, in Mobile VCE we structured it so that we had projects distributed around the universities [A021].

There were also some attempts to create integrated research within some of the programme areas. In Radio Environment, for example, it was claimed: 'The plan was drawn up right from the beginning to involve a lot more work between institutions . . . There are a lot of interdependencies between the packages of work' [A013]. Not all the programme areas were designed with this idea as a priority. The experience of the research associates (RAs) whom we interviewed was quite variable, some working closely with RAs in other institutions and others not. One commented that 'In the Terminals area there are five universities and each one of us is working on a completely different thing' [A016].

The extent to which RAs communicated with one another provides one perspective on the degree to which the VCE was acting as a virtual centre.

However, in response to a question about such communication, RAs tended to distinguish implicitly between the communication required by interdependencies in the research project, and communication arising from shared interests. In describing the first sort of communication, interviewees often cited the reason for communication. For example, one said: 'At present I'm working in a fairly isolated fashion. Last year I was working in collaboration on a programming task with researchers from Surrey and Kings' [A030/G1]. And another commented: 'We have to work closely on some aspects. For example, today I received five different e-mails concerning building layouts and drawings' [A015].

On the other hand, several RAs commented on the discussions that arise from casual meetings. As one put it: 'You have discussions with people at various meetings over coffee and that' [A030/G1]. Another mentioned the value of meeting academics in related areas at the steering committee meetings – '. . . when it turns out during the meetings that there has been some work done that is somewhere related to our work' [A020]. In this case, the meeting lead to an exchange of reports and documentation. In another case, a more social habit of communication followed an earlier research interdependency: 'At the beginning we were working together, until January. But we communicate whenever I think something is relevant to his work. We give each other references and books' [A024].

While the experience of Mobile VCE as an integrated research centre was variable among the RAs, nonetheless it does seem to have provided some support for informal, as well as formal, links between different academic institutions. This appears to depend not simply on the programmes of meetings set up for determining and monitoring research progress, but also on the design of the research programme to include some inter-institutional 'packaging' of the research.

THE INDUSTRIAL CONTEXT

At the time of the setting up of Mobile VCE, the mobile communications market was just beginning to take off, thanks partly to the success of the GSM standard (see below) for second generation digital mobile phones. At the same time, work was under way to institute the next (third generation) standard, and this provided a broad context for the research programme of Mobile VCE. In this section, we discuss how the industrial context may have contributed to the success of Mobile VCE in recruiting and maintaining a sufficiently large industrial membership to support its research programme.

A Brief History of Standards

The first generation of mobile telephones was analogue, which confined its use to voice communications. In addition, a variety of incompatible cellular systems in different parts of the world meant that users could not 'roam' from one country to another – a particular problem with the European Community attempting to open up trade links. The development of GSM, a European standard for digital cellular systems, began in 1982 in a study group called Group Spécial Mobile[14] set up by the Conference of European Posts and Telegraphs (CEPT). The responsibility for developing GSM was taken over by the European Telecommunications Standards Institute (ETSI) in 1989, first phase recommendations were published in 1990, and commercial operation of GSM networks began in Europe in 1991. In the UK, Vodafone launched a GSM network in 1991, One2One in 1993, Orange in April 1994 and BT/Cellnet in July 1994. The GSM standard was adopted beyond Europe. By the beginning of 1995, GSM networks were operational or planned in 60 countries in Europe, the Middle East, the Far East, Australia, Africa and South America.[15] By the late 1990s GSM was 'the standard of choice' in 118 countries and 40 per cent of the world's users.[16] In the USA, where the development of second generation standards involved a more *laissez-faire* attitude, four different standards were implemented, each backed by powerful industrial lobbies; and Japan had its own second generation standard, PDC (Personal Digital Cellular).

The success of GSM is often credited with creating a global market for mobile phones. As one interviewee put it:

> The thing that's made [the market] huge has been the adoption of a common standard across Europe and across much of the world, GSM. That's given the economies of scale that were necessary to get prices down and to give good return on the investment of companies who make the product. Before that we had a fragmented market, different systems in different countries, and it never really took off strongly [C001].

European companies in particular benefited from the market, which meant that they had cash to invest in longer-term research when invited to join Mobile VCE. In addition, work at Mobile VCE helped inform the development of third generation standards. The VCE's director, Tony Warwick, remarked:

> When we embarked on the programme we expected that all the standards for the next generation would be complete during the course of the research work. It turns out that for various reason that was delayed, and this research had more immediate impact in informing those decisions than we could have predicted.

GSM networks have the capability to transmit data other than speech but in practice are still mainly used for voice communication. Third generation systems offer increased capacity and much higher data transfer rates and are expected to extend the range of services available over the mobile phone network to include multi-media, and internet access. As *GSM World*, the web magazine of the GSM Association, puts it in its responses to frequently asked questions (FAQs) on third generation systems: 'Video on demand, high speed multimedia and mobile Internet access are just a few of the possibilities for users in the future.'

UMTS (Universal Mobile Telecommunications System) is a third generation standard, building on GSM, and part of the IMT-2000 third generation family of standards being developed by the ITU (International Telecommunications Union). The other enabling development for the third generation is the licensing of increased bandwidths, needed to support multi-media services, by the various national agencies. Finland awarded the first 3G licences in March 1999, and Germany, the UK, Austria and the Netherlands have all awarded licences. UMTS services became available in Japan in 2001, with pilots under way in the UK and elswhere. Issues related to the convergence of mobile telephony, the internet and multi-media services provide the context for Mobile VCE's Core Two programme, which started in spring 2000.

The Industrial Network

That Mobile VCE was successful in enlisting and retaining industrial members was significant for its continued existence. Not only did it enrol a sufficient number of members to fund its projected programme, but it also included most of the major companies in the mobile communications field.[17] However, it failed to attract any SMEs as Associate Members. In this section we explore some of the reasons given by industrial members for joining Mobile VCE and, more generally, consider reasons for the consortium's success in recruiting companies. We also draw some comparisons with the case of Digital VCE, which was set up at about the same time with a similar structure (though without Foresight Challenge funding) in the field of multi-media and digital signal processing, but which failed to attract the same number of industrial members.

Several of those interviewed mentioned previous experience of collaboration as an important precedent for Mobile VCE. The Radio Frequency Engineers training project was cited as bringing together the three original universities and some of the companies[18] involved in drawing up the bid for Mobile VCE; among other things, familiarity with the work of the universities was said to give companies confidence in the potential of their research. As discussed in the previous section, the industrial members of Mobile VCE also had experience collaborating in the discussions leading to the development of the European second generation standard, GSM. This was sometimes cited by members of Mobile VCE as one reason for the willingness of members of the mobile communications industry to work together. However, GSM was under development for nearly 10 years, and those involved would not necessarily repeat the experience. One industrial representative commented:

> Many organisations in Europe, with hindsight, look back on the American policy and say they did us a favour . . . but GSM did take an enormously long time to specify, much longer than it should have done. While GSM was an enormous success, there are no plans to repeat that process, because it did take too long. Every single bit of it was thrashed out in committees [C001].

On this view, the major lesson was the need to avoid the protracted defence of individual corporate interests. An academic member also remarked:

> There probably will never be a repetition [of the collaboration to build a GSM standard], because what is happening now is that we are looking to have an environment in which different standards can co-exist, interact and share infrastructure. The technology is advancing to the point where that is feasible; it wasn't when GSM was conceived [A021].

In addition to the experience of collaboration, standards may also be important in the mobile communications industry in a way that is not usual in other industries. One interviewee remarked on the contrast between mobile communications and the requirements of broadcasting transmissions relevant to the work of Digital VCE: 'In broadcast transmissions there are technical differences between the way in which satellite and terrestrial and cable will handle the detailed coding because they are addressing different problems; but they do not interact' [COO2].

The requirements of interactive communication are even more important in mobile telephony than in computing systems, and may be a reason for the perceived need to agree standards rather than allow a *de facto* standard to emerge (as happened for example in personal computers with Microsoft DOS and then Windows).

In discussing reasons for joining Mobile VCE, many industrial members pointed to its cost effectiveness as a means of gaining access to advanced research. Although they usually only participate actively in one area at most, all industrial members have access to the research results and the right to license products from all four core programme areas. The cost effectiveness is itself partly a result of the numbers involved in the project; as was discovered by members of Digital VCE, fewer members means a smaller research programme. The achievement of critical mass was important not only in respect to value for money, but also in persuading companies who saw that their competitors were all members and so they 'could not afford not to join'. Some interviewees, in citing cost effectiveness, also pointed out that market growth meant there was no problem in paying the subscription. However, one representative of a large multinational observed that he was required to justify membership on a yearly basis: 'It's to be compared with all the other opportunities for what we call "externalisation investment" globally' [B010]. The same speaker, asked to speculate on what might make his company leave Mobile VCE, summed up most of the reasons given by companies for membership:

> If that came about, it would be obviously in terms of value for money, if the fees are forced to increase significantly, perhaps through lack of membership; the chosen subject areas for future direction; and the efficiency with which it is operated – should it lose the support of some key players then that would also influence the situation.

To summarise, suggested reasons for the success of Mobile VCE in establishing an industrial consortium include:

- Experience of collaboration – RF Engineers Education Initiative
- GSM standard discussions
- Importance of standards
- Growing markets in mobile communications
- Cash available for research
- Vision of future
- Cost effectiveness/critical mass
- Return on investment
- Matching competitors.

Digital VCE, by comparison, was attempting to create a network in an industry that not only had no previous experience of collaboration, but did not necessarily identify other potential members as competitors or as having complementary research interests.

EVALUATION: CRITERIA AND PRACTICES

In the previous section, we explored a variety of reasons that might be given for the success of Mobile VCE in attracting and maintaining its industrial membership. This achievement also provides a measure of success in the evaluation of the VCE itself. However, there are many other questions to be raised about the effectiveness and the ultimate value of the VCE as a collaborative funding mechanism, as a mechanism for supporting excellence in research, and as a contribution towards promoting progress and growth, both from a national and a European perspective. This section provides a brief survey of some of the different criteria available for evaluating the VCE, considered first in the context of different institutionalised practices of evaluation, and second in relation to what we earlier called the policy discourse.

We referred to the horizontal reporting system practised within Mobile VCE; this involves the deployment of criteria such as research progress and the timely delivery of reports. In addition, industrial members are accountable to their own companies for their continued membership of Mobile VCE, which they again assess in terms such as research quality and exploitable intellectual property. In this context, the cost effectiveness of membership of the VCE is a significant criterion of evaluation. Academic members are also accountable to their own institutions and, while the winning of research money itself acts as an important justification of their membership, they may also need to show that it is useful in terms of research quality and as a source of journal papers and other indicators recognised in academic assessment,[19] and in particular in the Research Assessment Exercise (RAE). Other institutionalised practices of evaluation that may be relevant include those that may be carried out by the DTI/OST and the EPSRC, both of which customarily carry out formal evaluations. Table 3.1 indicates the way in which different criteria of success may be deployed in different institutional contexts. It is not intended to be an exhaustive list, but merely to illustrate how different groups do not necessarily share common terms of judgement.

The above terms of evaluation relate to formal or semi-formal practices of accountability which are self-evidently associated with different institutional or organisational contexts. However, although they are relatively stable, they are likely to change over time and are informed by the policy discourse. The policy discourse also opens up a broader set of terms of evaluation which are open to dispute and debate. For example the idea of 'excellence' in research and other areas has become central to the justification of the university (see Readings 1996), and the value of 'research excellence' is common to most of the above groups, but may be adduced in different ways for different communities. In this context, evaluation of the VCE involves questioning not only how well it works in its own terms (that is, attracting industrial membership, producing world-class research), but also how it contributes to broader political aims of generating social and economic benefit.

Table 3.1 Common terms of accountability in different institutional contexts

Institutional context	*Terms of accountability*
Academic Institution	Contributes research funds Generates journal papers (RAE value)
Company	Cost effective Ultimate market relevance
Research Council	Satisfies peer review (Delivers to timetable)[20]
DTI	Involves industry Academic/industrial links
Mobile VCE	Patentability Delivers to timetable

CONCLUSION

In the introduction we raised a number of questions concerning the interests of policy actors in Mobile VCE, referring both to the role of policy makers in initiating Mobile VCE and to the effectiveness of VCEs as a policy tool. While Mobile VCE undoubtedly 'worked' according to the terms in which it was set up, and under a number of different criteria as discussed in the previous section, this does not necessarily mean that VCEs are a useful policy tool. In conclusion, we summarise some of the specific issues relating to VCEs as a policy tool.

The virtual research network in Mobile VCE acted to encourage inter-institutional work among research associates, as well as to create links between researchers and the industrial representatives involved in research management activities. The political aims of creating a virtual centre included the creation of a centre of excellence (by bringing together isolated research teams) and the encouragement of industrial-academic collaboration through providing industry with access to a wide range of the best academic work in relevant fields. Both these aims are perhaps better served by a VCE than by a LINK-style of programme. A VCE, it might be said, brings together two networks (the academic and the industrial), not merely individual institutions. In this respect, Mobile VCE might stand as an example of one way of implementing the virtual centres envisaged in the European Commission's document *Towards a European Research Area* (CEC 2000, para. 1.1):

'The forms of teleworking which electronic networks permit make it possible to create real "virtual centres of excellence", in particular multidisciplinary and involving universities and companies.'

However the case of Mobile VCE also suggests that the technological infrastructure (forms of teleworking) are not necessarily as important in the building of a VCE as challenging institutional boundaries in the design of the research project.

The VCE mechanism exemplified by Mobile and Digital VCEs may also be contrasted to another slightly different type of VCE, sometimes described as the 'virtual shopping mall'. This approach is exemplified by some of the VCEs now being set up in the European Union, such as the Virtual Institute for Research in Official Statistics (VIROS), which claims:

> VIROS can be viewed as an implementation of a *Mall* or *Shopping Centre* metaphor. The idea is to have a system as decentralised as possible, every participating organisation remaining entirely responsible for its contribution; Eurostat would act as a central co-ordinator, ensuring that the individual elements appear integrated into a coherent, richer set.[21]

On the shopping mall metaphor, the actors are brought together only for accessibility to clients. The appropriate metaphor for the style of VCE exemplified by Mobile VCE would be a *virtual research centre*, in which there are common programmes and both formal and informal links between researchers and other actors, and which also provides a single site of access for clients.

Although Mobile VCE is usually judged a success[22], the VCE mechanism has not been reproduced beyond Mobile and Digital VCEs. One of the problems demonstrated by Digital VCE is the difficulty of identifying appropriate strategic sectors where a critical mass of industrial interests can be enrolled in support of a virtual research centre. However, other industrial sectors might well be appropriate. The main problem for the future of the mechanism seems to lie in the broader selection environment, that is, the political priorities of the various funding agencies. For example, the VCE mechanism, with its emphasis on inter-institutional collaboration, may be criticised for not encouraging competition between research teams; or the continued use of public funds in an industrially funded project may be questioned;[23] and while European funding priorities emphasise support for industrial development as the output of collaborative research projects, the VCE mechanism emphasises research that is prior to marketable products. Nonetheless, the VCE mechanism provides an original approach both to industrial-academic collaboration and the creation of a virtual research network.

NOTES

1. The case study is primarily based on interviews and a reading of texts relevant to the setting up of Mobile VCE. We carried out 30 interviews with academic and industrial members of Mobile VCE and representatives of some other relevant organisations, including the DTI, EPSRC and Digital VCE. Quotations from these interviews are referenced by an anonymised coding system.
2. At the time of completing the case study (October 2000), the industrial members of Mobile VCE were: BTCellnet, Dolphin, Ericsson, Fujitsu, Inmarsat, ITC, Lucent Technologies, Motorola, NEC, Nokia, Nortel Networks, NTL, One2One, Orange, Panasonic, Philips, Racal, The Radiocommunications Agency, Simoco, Texas Instruments, Toshiba and Vodafone.
3. This report was presented to ACARD by 1984, but not published until 1986.
4. The Foresight Steering Committee Report (OST 1995a, para. 3.2), for example, commented 'The Foresight processes ... have not been about picking winners but about generating a flexible, well-informed science, engineering and technology base, able to respond rapidly to our needs in future.'
5. The *Competitiveness* White Papers were published annually from 1994 to 1996 (see DTI 1994; 1995; 1996a).
6. The White Paper estimated the annual expenditure as about £6 billion. See para 1.18.
7. Phase I of Foresight lasted from 1993 to 1995 when the steering group and the first round panels reported (OST 1995a). Phase II included some reorganisation of the panels, including the merging of the Communications panel and the IT and Electronics panel into a single panel called IT, Electronics and Communication. Phase II ran from 1995 to April 1997. Phase III, which began in 1997, was intended to spread awareness of Foresight throughout UK business (see POST 1997, para. 4).
8. Foresight Challenge was one of the first major instruments of implementation of Foresight. It was launched in May 1995 with £40 million to spend over three years.
9. The Foresight Challenge proposal was the basis of the prospectus (DTI 1996b) which was circulated to companies interested in membership.
10. The 'shares' which companies buy to join the VCE are membership subscriptions.
11. £25,000 was the original price per share; it has since been increased to £30,000.
12. One academic commented: 'The Mobile VCE feels much more like working on a company contract. Company contracts are tightly managed, they have deadlines that you have to meet, working night-times and week-ends.'
13. One academic member commented that he was disappointed at the lack of research collaboration: 'In principle there is nothing to stop them doing pieces of work. They never have. It exposes them too much' (Interview A013). One reason given by an industrial member was that: 'We are not really a research organisation, we are a development organisation designing mobile phones, so we were better off paying somebody else to do some "blue-sky" for us' (Interview B006).
14. Scourias (1997) cites this name as the origin of the acronym GSM. *The Economist* (9 October 1999) suggests that it stands for Global System for Mobile Communication.
15. Scourias (1997).
16. *Economist*, 9 October 1999, p. 14.
17. This judgement was generally agreed by interview participants. As of late 1999, when we completed our interviews, the major non-member (cited by several interviewees) was the German company Siemens.
18. According to the Prospectus (DTI 1996b, p. 10) eight companies were involved in the RFE programme.
19. This criterion has not been applicable in the case of Mobile VCE, because EPSRC had only a financial management role; it has, however, organised peer review of the VCE's bid for renewed funding.

20. This criterion has not been applicable in the case of Mobile VCE, because the EPSRC had only a financial management role; it has however organised peer review of the VCE's bid for Core Two.
21. This description of VIROS is given on its home web page: http://europa.eu.int/comm/eurostat/research/viros
22. In terms of the usual sort of indicators, the first three-year programme achieved 20 patents, and nearly 100 technical papers. In a letter to *Research Fortnight* (13 September 2000), two board members of Mobile VCE (Barry Evans of the University of Surrey and Chairman Keith Baughan of Nokia) also stressed the academic-industrial collaboration, and achievements in training, spin-off contracts and 'culture change'.
23. See remarks by David Clark of EPSRC in *Research Fortnight*, 21 June 2000, p. 6.

REFERENCES

ACARD (1986), *Exploitable Areas of Science,* London: HMSO.

CEC (2000), *Towards a European Research Area*, Com. (2000) 6, Brussels: CEC.

DTI (1994), *Competitiveness, Helping Business to Win,* London: HMSO.

DTI (1995), *Competitiveness, Forging Ahead,* London: HMSO.

DTI (1996a), *Competitiveness, Creating the Enterprise Centre,* London: HMSO.

DTI (1996b), *Virtual Centre of Excellence in Mobile and Personal Communications,* London: DTI.

Economist (1999), Survey on Telecommunications, London: *The Economist*, 9 October 1999.

Faulkner, W. and J. Senker (1995), *Knowledge Frontiers.* Oxford: Clarendon Press and New York: Oxford University Press.

Gibbons, M., C. Limoges, H. Nowotny, S. Schwartzman, P. Scott and M. Trow (1994), *The New Production of Knowledge,* London and Thousand Oaks, CA: Sage.

Henkel, M., S. Hanney, M. Kogan, J. Vaux and D. von Walden Laing (2000), *Academic Responses to the UK Foresight Programme,* unpublished Report for the Nuffield Foundation, Uxbridge, CEPPP, Brunel University.

Leydesdorff, L. and H. Etkowitz (1996), 'The future location of research: A triple helix of university-industry-government relations II' (Theme paper for a conference in New York City, January 1998), *EASST Review*, **15** (4 December 1996), 20–5.

Martin, B.R. and J. Irvine (1989), *Research Foresight: Priority Setting in Science,* London and New York: Pinter.

OST (1993), *Realising Our Potential – A Strategy for Science, Engineering and Technology*, Cmd 2250. London: HMSO.

OST (1995a), *Report of the Steering Group of the Technology Foresight Programme*, London: DTI.

OST (1995b), *Report of the Steering Group of the Technology Foresight Programme. Number 6, Communications,* London: DTI.

POST (1994), *UK Technology Foresight,* London: POST.

POST (1997), *Science Shaping the Future? – Technology Foresight and its Impacts,* London: POST.

Readings, B. (1996), *The University in Ruins,* Cambridge, MA: Harvard University Press.

Scourias, J. (1997), 'Overview of the GSM cellular system. Extended Abstract', jscouria@www.shoshin.uwaterloo.ca

4. Innovation Networks in the Biotechnology-Based Sectors

Andreas Pyka and Paolo Saviotti

INTRODUCTION

Innovation networks are a relatively new phenomenon which have emerged in a significant way only since the beginning of the 1980s. Not only was this new phenomenon not predicted by economic theories but also its existence was considered an exception. The market and hierarchical organisations were considered to be the only stable and efficient forms of industrial organisation. Networks were expected to have only a temporary existence or to survive in special niches. As often happens, reality has taken science by surprise and the number of collaborative inter-institutional networks has steadily grown since the 1980s (European Commission 1997). There is then a need to modify existing theories of industrial organisation in order to explain the existence and the features of collaboration networks. The expression inter-institutional collaboration networks has been used before because the typical members of these networks are not only firms. Public research institutes or government departments participate quite often in these networks. Furthermore, in this chapter we are only going to be concerned with innovation networks; that is, with networks the main objective of which is to create and adopt innovations. This is not the only type of network, but it is the dominant one.

In most of the research about networks, the increased rate of creation of new knowledge and the shortening of the life cycles of products are two of the main factors associated with the existence of networks. Thus, mechanisms of knowledge creation and utilisation seem to be playing a very important role in the creation of networks. Networks can be considered a component of the emerging knowledge-based society, in which knowledge is expected to become the crucial factor leading to economic growth and to competitiveness. In a knowledge-based society not only will the quantity of knowledge used be greater but its mechanisms of creation and utilisation will change. According to Gibbons et al. (1994) a new mode of knowledge generation and utilisation, called Mode 2, is emerging in addition to the traditional one called Mode 1. While in the latter the creation and utilisation are clearly separated both chronologically and institutionally, in the former there is a continuous interaction between the two processes, which leads to the need for different institutional and organisational forms. Networks could, then, be a form of industrial organisation appropriate to a knowledge-based society.

Biotechnology is one of the fields at the forefront of the creation of a knowledge-based society. This seems somewhat paradoxical, since it could be maintained that biotechnology is one of the oldest technologies used by mankind. Beer and yoghurt making constitute two typical examples. However, modern biotechnology has been substantially changed by the advent of molecular biology, a new discipline which was founded in the 1930s based on the attempt to apply to biology the methods of physics. In the mid-1970s two discoveries, recombinant DNA and monoclonal antibodies, transformed a scientific discipline with a brilliant if distant future into a seedbed of industrial applications. Accordingly, some authors now call this latest vintage of biotechnology 'third generation', to distinguish it from the completely empirical first generation and from the second generation, which began with the production of antibiotics. Second generation biotechnology used scientific methods but it did not have the knowledge required to change the genetic make-up of organisms. Such knowledge was only provided in a systematic way by molecular biology. In the mid-1970s very few research institutions did research in molecular biology and they were mostly in the USA. The industrial firms that in principle could have exploited molecular biology did not have a knowledge base or an absorptive capacity for it. Their competencies and knowledge bases were concentrated in more traditional disciplines such as organic chemistry or microbiology. In fact, this lack of knowledge hampered firms' recognition of the opportunities that could have been offered them by molecular biology.

Biotechnology is not an industrial sector but a scientific field underlying a number of industrial sectors (pharmaceutical, agriculture, food, environment, etc.), here called the biotechnology-based sectors. Industrial applications of biotechnology are highly dependent on new scientific developments, even on those that are the result of basic research. Although the lead times between the discovery of new knowledge and its final embodiment in new products may be very long, the time between the creation of new knowledge and the funding of industrial research aimed at its applications is in general very short. Basic research is not exclusively confined to public research institutions, but it is also carried out by firms. Thus, both for what concerns its intensity of knowledge utilisation and for the mechanisms employed, biotechnology seems to be a very good example of industrial organisation in a knowledge-based society. Of course, the conclusions reached in this chapter will depend on the specificity of biotechnology, but they will also have some general significance for the analysis of a knowledge-based society.

The earliest analyses of networks of collaboration pointed to the possibility that they are only a temporary form of industrial organisation. Such temporary character could be the result of discontinuities in knowledge generation; for example, of the emergence of a new technological paradigm. It was argued that large diversified firms (LDFs) were committed to the old paradigm, in which all their competencies were concentrated, and that they could not easily internalise the new knowledge. Alternatively, LDFs did not have the absorptive capacity required to internalise the new paradigm and they were not capable of constructing it rapidly. A new type of industrial actor, small high-technology firms, emerged to bridge the institutional gap between public research institutions and LDFs. In the specific case of biotechnology such firms were called dedicated biotechnology firms (DBFs). DBFs were expected to act as intermediaries between LDFs and public research institutions. For our purposes, the DBFs performing this role will be called translators. In the course of time by collaborating with DBFs and with public research institutions, LDFs could construct a knowledge base and an absorptive capacity in biotechnology. Once this happened the role of DBFs would have become redundant and industrial organisation would return to the traditional dichotomy between the market and hierarchical organisations.

As previously pointed out, the rate of creation of inter-institutional collaborative networks has been steadily increasing throughout the 1980s and 1990s. Thus, either LDFs have not internalised the new paradigm constituted by biotechnology or a new role for DBFs has emerged in innovation networks. The analysis of this problem constitutes one of the main objectives of this chapter. Here, a second role for DBFs and thus for networks is discussed. By the end of the 1980s, LDFs in a number of industrial sectors had acquired a knowledge base in molecular biology (see, for example, Grabowski and Vernon 1994) and yet they continued to enter into collaborative agreements with DBFs. We hypothesise a second role for DBFs, linked to the extremely rapid rate of creation of new knowledge. Even if LDFs have acquired an absorption capacity for it, the sheer rate of advance is such that no LDF could keep up with it all. LDFs might thus use agreements with DBFs within networks in order to keep abreast of new developments that could turn out to have important economic applications. The alternative course of action for an LDF would be to invest in research in the same fields of biotechnology. However, with a very high rate of growth of knowledge this strategy would involve a very heavy, irreversible and risky commitment. The collaboration with DBFs constitutes a more flexible and reversible strategy. It is to be observed that this role does not involve a qualitative difference in the ability of

LDFs to understand molecular biology, but only the attempt to reach a better trade-off between readiness for action if promising developments were to emerge in new subsets of the biotechnology knowledge space and the sunk costs that need to be faced in order to keep these windows open. Furthermore, it must be remembered that the competitive advantage of LDFs is not constituted by their ability to understand new knowledge but by their capacity to combine the different competencies and complementary assets required to produce a final product. DBFs playing this second role will be called explorers.

In what follows we analyse biotechnology innovation networks with special emphasis on two types of actors, LDFs and DBFs. We compare the advantages and disadvantages for firms of going alone strategies and of networking strategies, taking into account the environmental factors influencing the formation of links between actors. Network formation will be shown to display a dynamics going beyond a first wave of networks and leading to a reorganisation of the partnerships involved.

THE SIMULATION MODEL

Before we explain the basic structure of our model, some remarks with respect to the methodological framework we are adopting are in order. In particular, we are going to use the methodology of the so-called history friendly models, recently introduced by Malerba et al. (1999). History friendly models are designed to capture, in a stylised form, the mechanisms and factors affecting industry evolution, technological advance and institutional change detected by empirical scholars of industrial economics, technological change, business organisation and strategy, and other social scientists. Thus, history friendly models can be considered the natural extension to modelling of qualitative and appreciative theories.

Obviously, even in an evolutionary approach simulation models have to introduce a certain degree of abstraction and cannot reflect reality in all its complexity. The mechanisms built into the formal model have to be transparent enough, so that the analyst can figure out what are the causes of the observed effects. Therefore, in the first step of our modelling effort we have to carefully single out the relevant actors, bring together variables which are effective in the same direction and combine important developments and possibilities of action. Nevertheless, adopting the approach of evolutionary economics allows us to emphasise crucial features of innovation processes, such as non-linear dynamics, heterogeneity and true uncertainty, which are beyond the scope of traditional approaches.

In the following pages, the basic building blocks of our modelling framework are introduced. In particular, we focus on the way we present the different agents in our model, the way we capture innovation processes, what we consider to be the prerequisites and consequences of networking as well as the representation of the economic realm in our simulation.

The Representation of Agents in the Model

The agents we are explicitly considering in our model are Large Diversified Firms (LDFs) and Dedicated Biotechnology Firms (DBFs). They are described in terms of their competencies and capabilities. DBFs possess technological competencies while LDFs possess a mixture of economic and technological competencies.

Competencies

Technological competencies are considered to be those components of the knowledge base required for building up production and innovation capabilities in a specific technology. In other words, before firms are able to develop new marketable outputs they have to develop the respective biotechnological competencies. Furthermore, technological competencies alone are not sufficient to achieve economic success with a new product. Economic competencies are necessary in order to successfully produce and market a new commodity. Examples of these economic competencies are experience in clinical trials, distribution channels and so on. Clearly this representation is somewhat simplified. The full range of competencies required by firms to conceive, develop, produce and market new products is very large and heterogeneous. However, given that most DBFs at the beginning of their life cycle do not possess any economic competencies and that LDFs in the 1970s were generally unable to acquire the knowledge required to use modern biotechnology, the representation in terms of technological and economic competencies adequately describes the difference between our two main groups of agents. Moreover, we could consider technological competencies as the core competencies (Prahalad and Hamel 1990) of firms and economic competencies as a large part of the complementary assets (Teece 1986) required to produce and market a product.

The building up of technological B_i^t and economic competencies can be described in the following equations:

$$B_i^t = \frac{1}{1+\exp(const - NCOP_i^t \cdot t^{BIO})} \qquad (4.1)$$

$$EC_i^t = \frac{1}{1 + \exp(const - NCOP_i^t \cdot t^{ECO})} \tag{4.2}$$

B_i^t = technological competencies of firm i at time t;
EC_i^t = economic competencies of firm i at time t;
$NCOP_i^t$ = accumulated number of co-operations of firm i at time t;
$t^{BIO/ECO}$ = time spent in particular activity.

Figure 4.1 shows this function graphically for the case of technological competencies. In the early phases the building up of the knowledge base is a difficult process and progress is hard to achieve. However, after having developed a certain knowledge base it becomes easier to learn even more (threshold effect). Finally, marginal progress becomes progressively difficult as the knowledge frontier existing at a given time is approached. A function of this type implies variable returns to investment in the creation of new knowledge within a given field: very low at the beginning, positive and growing in the intermediate phase before diminishing returns set in as the potential of the new field has been exploited. The process of building up a knowledge base in biotechnology is supported by co-operative arrangements with firms who are already active in this field – an important part of the respective knowledge base is transferred by networking.

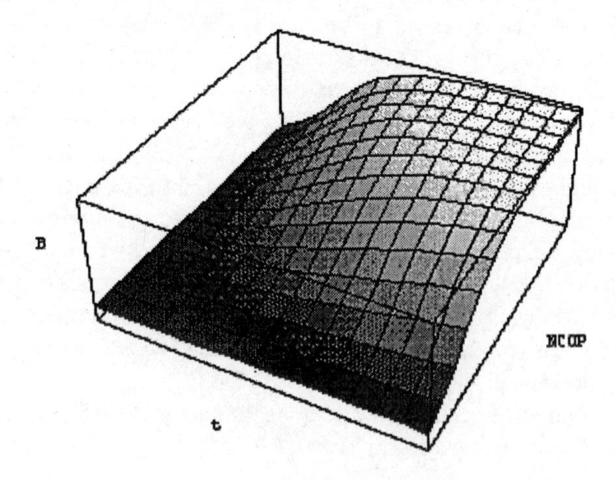

Figure 4.1 Building up of a knowledge base in biotechnology

Discrimination between LDFs and DBFs[1]

The two populations of firms which can be observed in the biotechnology-based industries can be distinguished on the basis of their relative technological and economic competencies. The first of these populations, the LDFs, are typified by the large established pharmaceutical firms. Until the end of the 1970s their research and development was mainly embedded in the paradigm of traditional organic chemistry. The emergence of the new biotechnological paradigm meant a 'competence-destroying technological progress' (Tushman and Anderson 1986) for the LDFs as most of their competencies were threatened by the new ones. In our simulation, this group of firms is represented in the starting distribution with well-developed economic competencies but with almost no technological competencies in biotechnology.

In the second population we find small start-up companies, often university spin-offs specialising in the biotechnology field. This group of firms, the DBFs, by their very nature have highly developed technological competencies, but almost no economic competencies. When they start their existence, DBFs depend on external funds for research and development. Accordingly, in our starting distribution they are represented just as having no economic competencies but highly developed technological competencies.

Venture capital firms and universities

In addition to these firms that we are explicitly taking into account, we also consider two further important groups of actors in our model: public research institutes or university laboratories and venture capital firms. In order to keep our model simple their behaviour is not explicitly analysed, but they are considered as an important component of the external environment of biotechnology firms. For example, in order to acquire the funds necessary to undertake R&D, a DBF can co-operate either with an LDF or with a venture capital firm; similarly, the co-operation of an LDF with a DBF or with a public research institute leads almost to the same consequences for the LDF, etc.

Capabilities

Drawing on their competencies, firms can accumulate technological capabilities in specific fields which allow them to explore further technological opportunities. The firms in our model act in an environment which continuously forces them to be engaged in R&D processes. Not to innovate means to fall behind in the competitive environment of biotechnology. In order to increase the probability of an innovation, firms accumulate technological capabilities in the course of time as the following equation demonstrates:

$$C_i^t = \sum_t r_i^t \tag{4.3}$$

C_i^t = capabilities of firm i at time t;
R_i^t = R&D net-investment of firm i at time t.

Together with the technological competencies B_i^t the technological capabilities determine the probability of an innovation Pr_i^t which is described as:

$$\mathrm{Pr}_i^t = 1 - \exp(-B_i^t \cdot C_i^t) \tag{4.4}$$

Pr_i^t = innovation probability of firm i at time t.

To consider the intrinsic uncertainty of innovation processes the innovation probability of a firm is matched every period with a Poisson-distributed random number whose mean value is asymptotically reached by Pr_i^t. A firm is successful in its innovative efforts only if the innovation probability Pr_i^t is above the random number.[2]

However, technological capabilities are not sufficient for the successful introduction of a new product. To do this a firm also needs to acquire economic capabilities E^t_i as well as the economic competencies EC^t_i, for example, in production, legal approval, marketing, distribution, etc. The economic capabilities are accumulated in the same way as the technological capabilities and are responsible for incremental innovation on new technological trajectories opened up by a product innovation.

R&D Decision Rules

The investment in R&D is no longer guided by an optimisation calculus, but by a routinised behaviour, as innovation goes hand in hand with true uncertainty (for example, Nelson and Winter 1982, p. 132). Firms adopt certain rules: for example, invest x per cent of your turnover in R&D, retain x per cent of your financial support in order to build up an own capital stock, etc. In the same way, the distribution amongst different activities (for example, between investing in the building up of technological or economic capabilities) is captured by referring to routines.

Networking

In order to carry out their innovation processes firms can choose different strategies. They can either decide to go-it-alone, which means not to draw on external knowledge sources and not to share their own new know-how with potential competitors, or they can decide to co-operate with other actors and build up collectively the new capabilities necessary for the introduction of a new commodity. Innovation networks emerge by this mutual co-operation, which gives rise to channels for knowledge flows between the firms participating in the network. In particular, we consider the evolution of innovation networks at three levels within the model: the environmental conditions favouring or inhibiting the growth of networks, the individual decisions of firms to co-operate or not, and a matching process bringing together firms willing to co-operate. This process creates a population of networks with its own dynamics. The formation of any network constitutes an act of birth or entry into the population. Conversely, the disappearance of a network constitutes an act of death or exit. The dynamics of birth and death of networks will be determined by the specific features of each network and by some features of the external environment. Accordingly, we can calculate the probabilities of birth (P'_B) and of death (P'_D) that will contribute to the net probability of network creation (P'_N).

Probability of birth

A number of environmental factors increase the probability of the birth of innovation networks. The growing complexity of innovation processes as well as a high degree of technological uncertainty play the most important role. Every time a firm is successful in introducing an innovation the number of knowledge fields *#KB* is assumed to grow. Given the complementary and combinatorial nature (see, for example, Staropoli 1998, p. 15) of biotechnology, the technological space Ω, defined as the number of possible combinations of knowledge fields, increases in a non-linear way as shown in the following equation (and see Figure 4.2):

$$\Omega = \frac{\#KB!}{2!\#(KB-2)!} \tag{4.5}$$

Ω = technological space;
#KB = number of different knowledge fields.

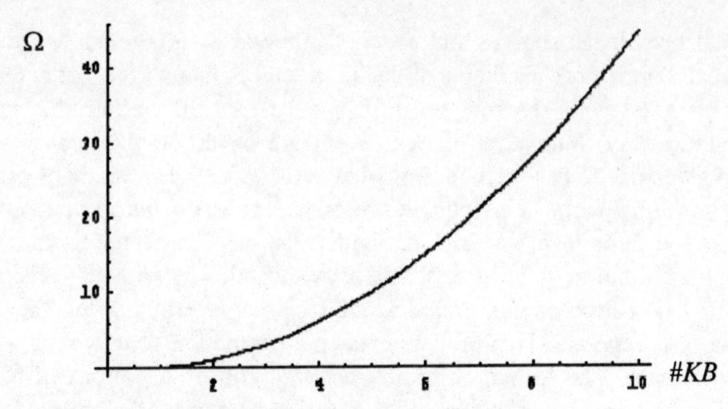

Figure 4.2 Increasing technological space

Especially in the early phases of a technological life cycle (*TLC*) this increasing complexity is combined with a high technological uncertainty because specific research techniques or heuristics –for example, how to handle this complexity – are not yet developed. In the model, the phase of a technological life cycle is approximated very roughly by the average age of the different commodities on the product markets.

Additionally, R&D networks are dependent on a number of core technologies or core/central actors (M_N), who play a crucial role in the establishment of the networks (see, for example, Saviotti 1998, pp. 36–7). In our model the population of LDFs is supposed to play this role. These factors influence the probability of birth of innovation networks P^t_B as summarised in the equation, the function form of which implies a sigmoid relationship (as shown in Figure 4.3).

$$P_t^B = \frac{1}{1+\exp\left(const - \frac{1}{TL}\cdot\frac{M_N^t}{N_N^t}\cdot\Omega\right)} \qquad (4.6)$$

P^t_B = probability of birth of innovation networks;
TL = age of technology life cycle;
M_N = number of core actors;
N^t_N = number of networking firms.

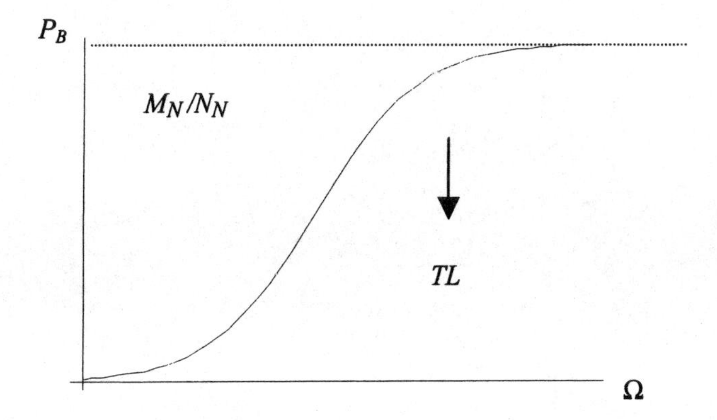

Figure 4.3 Probability of birth of innovation networks

In other words, the probability of birth of innovation networks increases with the complexity of the technological space and with the number of core actors, and it decreases with the age of the technological life cycle.

Probability of death

In addition to the previous network supporting effects, other influences decrease the probability of network formation, thus leading to network death. First, the degree of competition is crucial in this respect. In our model we use as an economic framework a heterogeneous oligopoly. We consider the degree of substitutability of the final products as a measure for the intensity of competition. We use the variance σ_a^t of the variables describing a firm's relative product quality a_{ij}^t as a measure of the product heterogeneity: the higher this variance, the lower is the competitive threat between the firms. Furthermore, we can expect demand saturation to decrease the rate of growth of the respective markets and thus the scope of co-operative R&D. In this phase of the industry life cycle, minor improvements of the technology could lead to considerable advantages for a single firm. To capture this influence, we draw again on the *TLC* and assume that in its later stages the rate of growth of demand is likely to decrease. Finally, the techno-economic performance of the k network members, again approximated by the relative quality compared to the average performance of all firms, is itself an indicator for the attractiveness of joining a network. In cases where the performance of network members is below the average performance of the whole firm population, the networking strategy significantly loses attractiveness. These factors in the probability of network death P^t_D are summarised as:

$$P_D^t = \alpha_1 \cdot \left(\overline{a}^t - \frac{1}{k} \sum_k a_k^t \right) \cdot \frac{1}{\sigma_a^t} \cdot TL \tag{4.7}$$

P_D^t = probability of network death;
$\overline{a}^t$ = average quality at time t;
$\sum_k a_k^t$ = average quality of co-operating firms;
σ_a^t = heterogeneity on product markets;
α_1 = weighting parameter.

The net probability P_N^t of network creation at any given time is determined by the balance of births and deaths. The value of P_N^t in our firm population determines the decision of firms to engage or not to engage in co-operation:

$$P_N^t = \frac{1 + P_B^t - P_D^t}{2} \tag{4.8}$$

P_N^t = net probability of innovation networks.

In cases where P_N^t is below 0.5, $2 \cdot (0.5 - P_N^t) \cdot 100$ per cent of firms previously engaged in co-operation turn away from co-operation, if P_N^t is above 0.5, $2 \cdot (P_N^t - 0.5) \cdot 100$ per cent of firms engage in further networking. Accordingly, the probability P_N^t determines the number of firms who decide to cooperate or not in every period t.

Networking decisions

Next, the firms have to decide whether they want to co-operate or not. Generally, two forms of co-operation are possible:

i) Co-operation focusing on complementary assets; that is, firms are induced to co-operate to acquire technological or economic competencies that they do not possess but that they judge crucial for their economic success.
ii) Co-operation focusing on general complementarities; that is, the bundling of R&D efforts in a specific direction and synergies; that is, detecting potentials for cross-fertilisation by the combination of different technological capabilities. It is to be noted that in this latter case, co-operating firms can have competencies with a greater degree of overlap than in case i). For example, it is possible to conceive a division of labour in which firms pursue similar objectives using similar competencies, but they collaborate

in order to speed up the innovation process and to spread the relative uncertainty over the network.

In the form of co-operation i) DBFs play the role of translators, while in form ii) they play the role of explorers. Consequently, the networking decision depends on the respective competencies and capabilities that firms have accumulated. For example, a small start-up DBF in its early phases is not able alone to raise funds for R&D and necessarily has to look for a partner in order to obtain funding. In the same way established LDFs which want to become active in the promising fields of biotechnology but have no internal technological competencies need collaboration partners experienced in these fields. On the other hand, firms with highly developed capabilities would not run the risk and share their knowledge with potential competitors in the stages immediately preceding the introduction of an innovation.

Matching process
Finally, we have to decide on the mechanism which brings together different firms willing to co-operate. Although different mechanisms are conceivable we think that a mechanism which could be labelled 'success-breeds-success' is best suited to our purposes. Success-breeds-success means that firms would tend to pick collaborators with the highest technological and/or economic capabilities. Here, we assume that firms are able to advertise their own capabilities and to value those of potential co-operating partners. This seems to be a realistic assumption, especially in the biotech industry, where firms are ranked on the basis of their technological performance, which is advertised by press announcements, publications, patents and even by the professional standing of the scientists hired by firms, including the Nobel prize winners present in their scientific committee.

Networking consequences
After having introduced the way firms get together in innovation networks we now have to focus on the consequences of networking. By entering into a collaboration, firms are exchanging their know-how. This means that firms can benefit from the efforts of other firms in order to build up their own capabilities.

Absorptive capacities
The extent to which a firm can benefit from the knowledge flow available by co-operation depends on its absorptive capacity (see Cohen and Levinthal 1989; Cantner and Pyka 1998). In turn, absorptive capacity is expected to increase with the firm's previous experience in co-operation. This is represented by the experience term δ_i^f, which describes the amount of external

competencies a firm is able to integrate – a kind of absorptive capacity in networking. This means that external knowledge is not easily integrated within the own knowledge stock, but certain prerequisites have to be fulfilled and a minimum amount of experience is necessary. This also means that the amount of knowledge which flows within the network is severely limited. The building up of the absorptive capacity is given in the following equation where we draw on a firm's experience in co-operation as an approximation:

$$\delta_i^t = \alpha_2 \cdot NCOP_i^t \tag{4.9}$$

δ_i^t = absorptive capacities of firm i at time t;
α_2 = weighting parameter.

Co-ordination costs

Co-operation also involves costs, which reduce R&D investments. These costs together with the prevailing environmental conditions determine the potential number of collaborations in the industry while the decision rule described above determines the form of co-operation chosen. Of course, not all firms are engaged in co-operative relationships with all other firms, so there have to be certain limits to a co-operative strategy. An important factor limiting the growth of networks is that of co-ordination costs, cr^t_i, which appear together with co-operative R&D. We assume these co-ordination costs to be constant and equal for every form of co-operation. These costs of co-operating with other firms decrease the budget for direct research r^t_i, since there is a trade-off between engagement in acquiring internal and external knowledge. The following equation shows this constraint:

$$R^t_i = r^t_i + COP_i^t \cdot cr^t_i \tag{4.10}$$

R^t_i = gross R&D budget of firm i at time t;
COP_i^t = number of co-operations a firm is engaged in at time t;
cr_i^t = co-ordination costs.

Therefore, in deciding whether or not to engage in co-operative R&D, firms also consider these co-ordination costs cr^t_i. For a firm engaged in several co-operative relationships, co-ordination costs amount to $COP_i^t \cdot cr^t_i$. They should not exceed a certain percentage η of the gross R&D budget R_i^t. Accordingly, the following equation gives the decision rule that has to be considered by a firm in addition to the decision as to the specific form of co-operation it prefers:

$$\begin{array}{lll} \text{if} & COP_i^t \cdot cr_i^t \geq \eta \cdot R_i^t & \text{then no further co-operation is intended} \\ \text{else} & COP_i^t \cdot cr_i^t < \eta \cdot R_i^t & \text{then new co-operations are possible.} \end{array} \tag{4.11}$$

FINANCIAL FLOWS

Start-up DBFs with missing economic competencies cannot finance their own R&D and are obliged to find a co-operation partner. In this case an LDF co-operating with a DBF is supposed to provide the required research funding. For a DBF co-operating with an LDF this means that its gross R&D budget R_i^t is financed in a part by the other firm's R&D budget. Of course, the DBF will also retain a certain percentage κ of the funds as profits $((1-\kappa)R_i^t = r_j^t)$ thereby acquiring means which in future allow it to undertake R&D more independently. In the case of a successful innovation, the intellectual property rights belong to the LDF which can start production of the final product.

Another possibility for DBFs to acquire R&D funds is to apply for venture capital, for which we assume an exogenous supply VC^t growing at a constant rate. Accordingly, the number of firms which can be financed by venture capital is $n^t_{VC} = VC^t/R$ (R = constant periodically paid amount of money). Furthermore, we assume a constant period t^{VC} in which the respective firms have access to venture capital. Access to venture capital is competitive. Amongst the firms applying only those which show the best record in biotechnological capabilities as well as in previous co-operations are likely to be funded.

Knowledge Flows

One of the most important advantages of participating in an innovation network is the access to channels of knowledge flow. External knowledge exerts an impact on the innovation probability function and depends on the amount of absorptive capacities, as well as on the technological capabilities of the co-operating firms. For a firm participating in an innovation network and collaborating with k other firms the innovation probability function becomes modified as demonstrated in the following equation:

$$\Pr_i^t = 1 - \exp[-(B_i^t \cdot C_i^t + \delta_i^t \cdot \sum_k C_k^t)] \qquad (4.12)$$

$\sum_k C_k^t$ = capabilities of k co-operation partners.

Thus, participating in an innovation network exerts a threefold influence: first, the research budget of a firm is reduced due to co-ordination costs and, in the case of a co-operation with a DBF, by the financial support of this firm. Second, absorptive capacities are positively influenced by entering into a new collaboration as the experience with integrating external knowledge is increasing. Finally, external knowledge becomes available via knowledge flows between the collaborating firms.

Competition Processes

The innovative activities of firms are undertaken in an economic environment which is characterised by a certain degree of competition. On the one hand, the firms offering products on the final market compete with each other in attracting demand. Also, those firms that aim to offer new technological knowledge compete in a particular way with other firms in acquiring the respective funds. Finally, firms that want to buy the respective knowledge also compete for the co-operation with the most attractive research laboratories.

The two levels of competition take place in two different markets: the market for final products and the market for knowledge. Of course, we know that markets for knowledge cannot exist due to their imperfections. However, the existence of DBFs, which very often though not always function as contract research organisations, implies that within particular circumstances such imperfections can be reduced to a level where a market for knowledge, although very imperfect, can exist. In fact, co-operation often exists between firms operating in different markets (for example, for final goods and for knowledge) and thus being in a complementary relationship. Of course, this does not exclude that firms operating in the same market can co-operate.

On the final markets, firms compete in terms of prices and quality which are, in a dynamic context, determined by their innovative success. Generally, one would expect that a successful innovator will be able to attract demand away from its competitors because consumers can choose between several goods. These substitution effects are due to price and quality changes which are the results of the following actions and reactions:

- Introducing a new product with improved quality characteristics creates additional demand allowing the innovator to charge higher prices.
- In the case of an introduction of a new product by two or more vertically integrated firms who co-operated in the R&D stages the increase in demand is divided between the involved firms.
- As a reaction to this quality-induced substitution effect, non-innovators in related markets lower their prices in order to keep the loss in demand as small as possible.
- Exploitation of technological opportunities of an already existing technology allows the respective innovator to reduce its price, thereby increasing the demand for its product;
- As a reaction, non-innovators could fight their loss in demand by also lowering their prices, thereby, however, reducing their profit margin.

Another form of competition takes place in finding the most attractive network partner which is described with the help of the notion success-breeds-success. Firms engaged in the search for a co-operation partner will match with those which show either the most developed technological or the most developed economic competencies.

By choosing a heterogeneous multi-product oligopoly (see Kuenne 1992; for an application in a simulation model, see Cantner and Pyka 1998) we allow for the relationships described above. Firms are offering their goods on a heterogeneous product market. By an innovation and the introduction of a new commodity on these markets the relative market share of the already existing goods gets eroded. By this, we also generate the endogenous incentives of the firms to engage in innovation, as they cannot survive in the long run by relying on their original established positions which are continuously threatened by the innovative actions of their competitors.

The Basic Structure of the Model

The following flow chart summarises the basic structure of our model. Starting with firms and industry characteristics of the previous round firms have to decide whether to go-it-alone or to co-operate. They are influenced by environmental conditions either favouring or inhibiting the growth of networks. After having found a co-operation partner in the matching process the firms enter the innovation stages which on the one hand influence the industry and firm characteristics, and on the other hand the market outcomes of the next round.

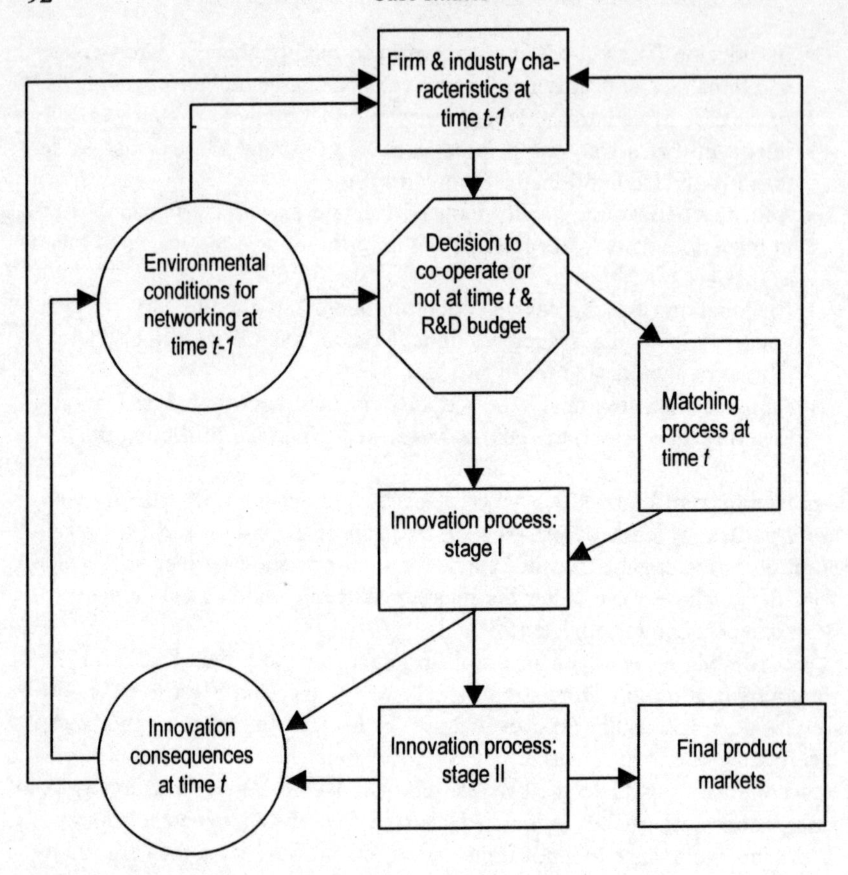

Figure 4.4 Flow chart

THE MODEL'S RESULTS

The first simulation experiments deal with a purely theoretical case where we are considering a population of 12 firms, four of them being LDFs and eight DBFs. This step seems to be necessary as an introduction to the model's results. Therefore, these first simulation results have to be seen as interim results demonstrating the basic functioning of our model as well as the plausibility of the implemented relationships and dynamics. The network-related results are then compared to real figures in a history friendly manner.

Before analysing the development of the network structure we begin with the environmental conditions and some figures describing the typical course

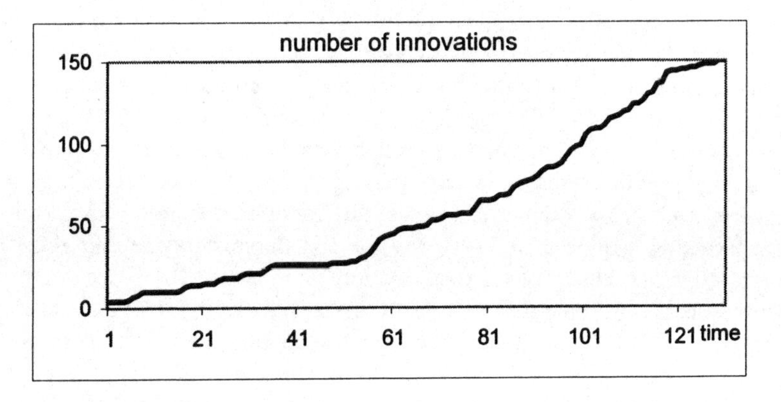

Figure 4.5 Number of innovations

of single firms. In Figure 4.5 we see how the number of successful innovations develops.

The first 50 periods are characterised by a slow introduction of innovations. During these periods firms are mainly occupied with building up the prerequisites to cope with technological progress. The rate of creation of innovations starts accelerating only after period 45. During this period firms build up the required technological competencies and experience in networking (absorptive capacities). The introduction of innovations accelerates even further around period 55, where nearly after every second period a new commodity appears on the markets

In Figure 4.6 we find the development of the average age of the industry life cycle approximated by the average age of the products offered. The aver-

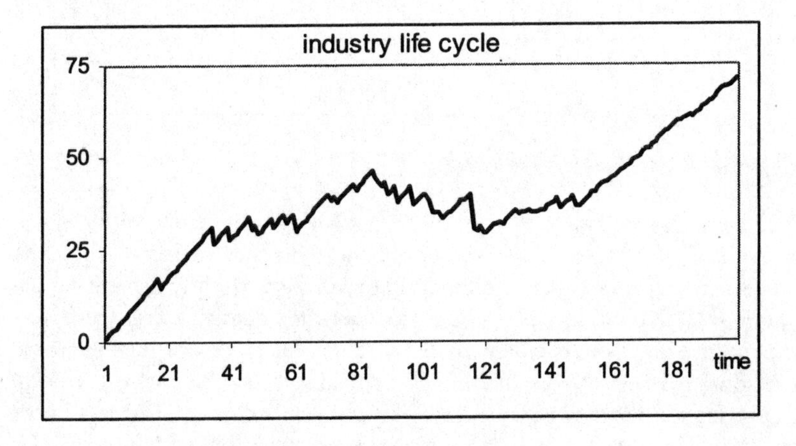

Figure 4.6 Average age of the industry life cycle

age age increases at a faster pace during the first 40 periods due to the slow rate of introduction of novelties. However, as innovative activity starts accelerating at the end of this period, the average age of products starts oscillating around a mean of 40. The ageing process is thus reduced and with it the negative impact on the incentives to collaborate. Later on, around period 165 the average age increases again which is caused by the co-existence of a larger variety of commodities introduced at different times. Here we find, again, the alternating sequence of low, growing and diminishing returns already found for the probability of network creation.

Both effects determine the environmental conditions for networking shown in Figure 4.7. During the first 40 periods the combination of an increasing age of the industry life cycle and a relatively low rate of introduction of innovations worsens the environmental conditions for networking. After period 60 the increasing rate of creation of innovations favours the growth of networks, which can mainly be traced back to an increasing technological space. This effect even outweighs the further ageing of the industry life cycle in later periods.

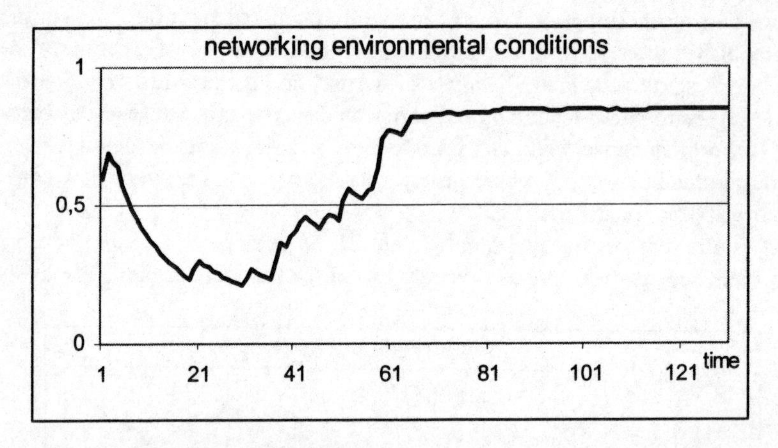

Figure 4.7 Environmental conditions

The corresponding development of the network density is shown in Figure 4.8. After a first increase in the density the dynamics of network growth come to a rest after around 10 periods and even starts slightly decreasing until period 45. However, after that period network density starts increasing again, until it begins to oscillate around a value which is twice the average at the beginning. This can be interpreted as evidence for the changed role of DBFs, which in the first periods find temporary collaboration partners in the population of LDFs. These collaborations are mainly oriented towards bridg-

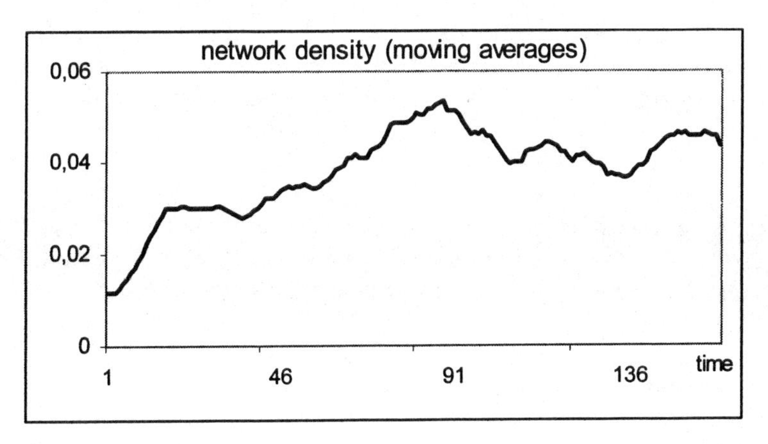

Figure 4.8 Network density

ing the gap between new biotechnologies and established industry. Later on, however, the DBFs are finally considered as an extension of internal R&D facilities, allowing LDFs to explore a wider opportunity space. Therefore, collaborations become more frequent and lasting in more advanced states of the industry evolution.

This interpretation of the changed role of DBFs is supported when we analyse the developments at the firm level. The process of building up technological competencies by four LDFs is shown in Figure 4.9. All of them begin with no technological competencies in biotechnology, but have to build them up by collaborating with firms specialising in these fields, namely DBFs. Two of the firms engage early in collaboration (thin curves) while the other

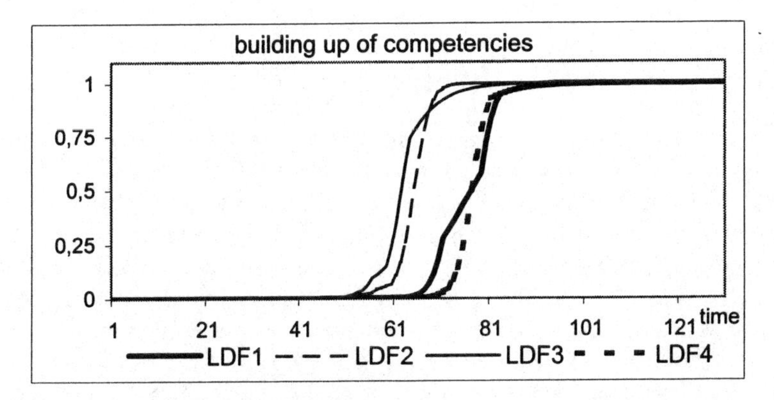

Figure 4.9 The building up of technological competencies by LDFs

two (bold curves) start only at a later stage. The two firms engaging early in collaboration improve more rapidly their competencies and accordingly reach sooner the second branch of the learning curve, with positive but decreasing rates. These firms develop considerable competencies in biotechnology more than 20 periods earlier than the two slow ones. We can observe that both for the fast and slow collaborating LDFs the shape of the learning curve is sigmoid and that the saturation level seems to be the same. From these results it seems that no penalty needs to be paid for late entry. While this result may depend on some features of our model that deserve further investigation, they are limited to learning and they do not take into account possible barriers of other types that might be created during the technological life cycle.

This building up of competencies has an immediate effect on the innovation probability, which is guided by the exponential relationship of Equation (4.12) with positive but decreasing rates. Figure 4.10 shows the development of innovation probabilities of three DBFs (thin lines) and three LDFs (bold lines) in the starting periods.

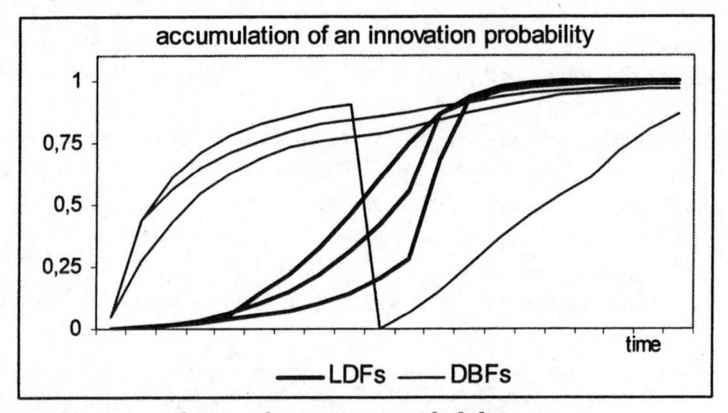

Figure 4.10 Accumulation of innovation probabilities

The DBFs are able to accumulate quickly their innovation probabilities depending on their success in acquiring resources for R&D. One of the three firms is even able to introduce a first innovation and to begin developing a second within the period shown in Figure 4.10. Compared to DBFs, the population of LDFs is confronted with severe difficulties in exploiting their first trajectories. They need a considerably longer time to build up their technological competencies to the level required to innovate. Of course, this varies depending on the LDFs' networking strategy. However, in general LDFs need more time to reach promising innovation capabilities. In the early period of biotechnology they depend on collaborations with firms from the population of DBFs in order to access the technological space offered by this new field.

Above we argued that the persistence of innovation networks in biotechnology-based industries could not be explained by means of only one role played by DBFs. Whereas in early stages the small technology-oriented firms play the role of translators, facilitating the absorption of the new technologies by LDFs, in later stages they become more emancipated as collaboration partners. This means that they do no longer serve solely as institutions transferring knowledge between academic and industrial research, but become explorers, allowing LDFs to investigate a broader technological portfolio in an increasing complex technological opportunity space. This changed role of the DBFs accordingly has to be observed also in the simulation as a development which endogenously takes place within our model's specification. In Figure 4.11 we therefore plot the specific composition of co-operative agreements.

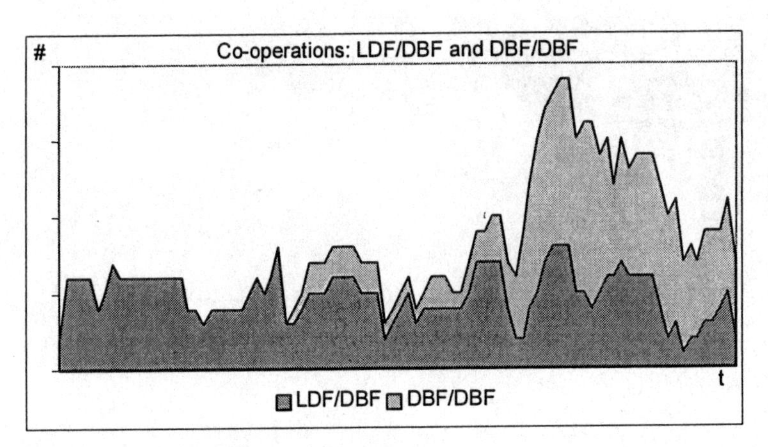

Figure 4.11 Composition of co-operations

In the first part of the period investigated only co-operative arrangements between LDFs and DBFs are found: DBFs are supposed to support LDFs in building up their biotechnology competencies; and, as a compensation for their R&D efforts, they are funded by LDFs. As soon as some DBFs start to earn their own money they also initiate further collaborations in which they are no longer playing the role of translators but that of explorers. In the simulation we find that these collaborations between DBFs become of increasing importance in later stages of the simulated time horizon. Now, the co-operative agreements aim at bundling know-how and joint exploration of technological opportunities. At the end of the period studied the number of agreements between DBFs is becoming comparable to that of agreements between LDFs and DBFs.

This changed role played by DBFs is also mirrored in the decisions made by firms with respect to their collaboration policy. In the model we distinguish between three strategies: the go-it-alone strategy chosen by firms who are either at the technological frontier and don't want to share know-how with followers or by firms which already are engaged in several co-operations. The second strategy aims at attracting research funds; this strategy is adopted by the DBFs in their early phases, when they enter the scene with highly developed technological competencies but they have no economic competencies. Finally, the third strategy aims at the integration of external knowledge in order to build up jointly in a network the capabilities necessary for the introduction of an innovation.

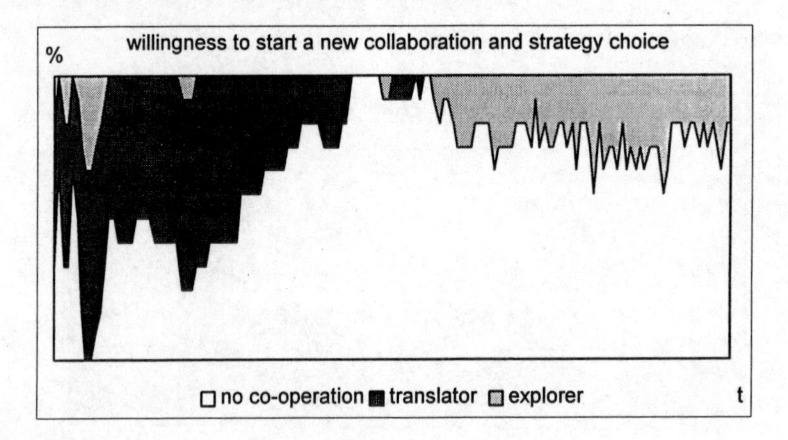

Figure 4.12 Strategy choices

Figure 4.12 shows the share of strategy choices the firms (LDFs and DBFs) make with respect to the specific form of co-operation they want to initiate. This decision, of course, is always influenced by the position of the firm, in particular depending on whether the firm is already engaged in one or more co-operations. Therefore, the white area representing the proportion of firms who don't want to start a new co-operation has to grow with the number of already existing networks because of the increasing co-ordination costs. In the early stages almost all firms wish to start new co-operative relationships according to the translator's type (black shaded area). With the growing diffusion of technological competencies within the population of LDFs and as some successful DBFs become vertically integrated producers, this decision shifts nearly exclusively to collaborative relationships following the explorer's type (grey shaded area). This also means that the new collaborative agreements to be started in later periods will almost only be of this latter type which is in line with the results shown in Figure 4.11.

COMPARING ARTIFICIAL AND REAL NETWORKS

We can compare our artificial world with developments in the real world. In order to get a first idea of the respective artificial data, four selected periods are freeze framed in Figure 4.13.

Ad 4.13a): In the starting periods co-operations are focused on acquiring complementary assets; that is, DBFs are looking for financially powerful partners whereas LDFs are looking for technologically interesting partners with core competencies in biotechnology. In this situation we end up with all firms in the population of DBFs collaborating with one or two partners out of the population of LDFs.

Ad 4.13b): In this period most of the early co-operations are terminated. This is caused mainly by two effects. On the one hand, a decreased network probability caused by an advanced age of the first technology life cycle leads to a cancelling of less successful collaborations. On the other hand, some collaborations have led to an innovation and are terminated afterwards. As there are new collaborations, a re-orientation with respect to the selection of partners has taken place. LDF2 still participates in a network with two DBFs and also LDF3 has increased co-operative engagement by now co-operating with two DBFs. The most significant change has taken place with respect to DBF1 which obviously was successful in becoming a vertically integrated supplier. This firm no longer collaborates with any LDF but instead has built up a network with three other DBFs (DBF2, DBF3, DBF4).

Ad 4.13c): When we look at the networking table of period 75 the situation has changed once more. Now we only find four collaborative agreements and none of the early co-operations is still in existence, however LDF4 and DBF2 are again collaborating in a network. This change is mainly caused by successful innovations as well as a still slow improvement of the environmental conditions for networking.

Ad 4.13d): In later stages (period 95) most of the indicators support the emergence of innovation networks and accordingly we find a dense network between LDFs and DBFs and between those DBFs who were successful in becoming vertically integrated producers (DBF1, DBF2, DBF3). Also, all of our four LDFs are engaged again in eight different collaborations supporting our hypothesis that the role of DBFs is changing from translators to explorers in the course of time making innovation networks a persistent phenomenon.

3	DBF_1	DBF_2	DBF_3	DBF_4	DBF_5	DBF_6	DBF_7	DBF_8
LDF_1	**1**	0	**1**	0	0	0	0	0
LDF_2	0	**1**	0	**1**	0	0	0	0
LDF_3	**1**	0	0	0	0	0	0	0
LDF_4	0	**1**	0	0	0	0	0	0
DBF_1	0	0	0	0	0	0	0	0
DBF_2	0	0	0	0	0	0	0	0
DBF_3	0	0	0	0	0	0	0	0
DBF_4	0	0	0	0	0	0	0	0
DBF_5	0	0	0	0	0	0	0	0
DBF_6	0	0	0	0	0	0	0	0
DBF_7	0	0	0	0	0	0	0	0
DBF_8	0	0	0	0	0	0	0	0

a)

75	DBF_1	DBF_2	DBF_3	DBF_4	DBF_5	DBF_6	DBF_7	DBF_8
LDF_1	0	0	0	0	0	0	0	0
LDF_2	0	**1**	0	0	0	0	0	0
LDF_3	0	0	0	0	0	0	0	0
LDF_4	0	**1**	0	**1**	0	0	0	0
DBF_1	0	0	0	0	0	0	0	0
DBF_2	0	0	0	0	0	0	0	0
DBF_3	0	0	0	0	**1**	0	0	0
DBF_4	0	0	0	0	0	0	0	0
DBF_5	0	0	0	0	0	0	0	0
DBF_6	0	0	0	0	0	0	0	0
DBF_7	0	0	0	0	0	0	0	0
DBF_8	0	0	0	0	0	0	0	0

c)

48	DBF_1	DBF_2	DBF_3	DBF_4	DBF_5	DBF_6	DBF_7	DBF_8
LDF_1	0	0	0	**1**	0	0	0	0
LDF_2	0	0	**1**	**1**	0	0	0	0
LDF_3	0	**1**	**1**	0	0	0	0	0
LDF_4	0	0	0	0	0	0	0	0
DBF_1	0	**1**	**1**	**1**	0	0	0	0
DBF_2	**1**	0	0	0	0	0	0	0
DBF_3	**1**	0	0	0	0	0	0	0
DBF_4	**1**	0	0	0	0	0	0	0
DBF_5	0	0	0	0	0	0	0	0
DBF_6	0	0	0	0	0	0	0	0
DBF_7	0	0	0	0	0	0	0	0
DBF_8	0	0	0	0	0	0	0	0

b)

95	DBF_1	DBF_2	DBF_3	DBF_4	DBF_5	DBF_6	DBF_7	DBF_8
LDF_1	0	0	**1**	0	0	**1**	0	0
LDF_2	0	0	0	0	**1**	0	**1**	0
LDF_3	0	0	0	0	**1**	**1**	**1**	0
LDF_4	0	**1**	0	**1**	0	0	0	0
DBF_1	0	**1**	0	0	0	0	0	0
DBF_2	**1**	0	**1**	0	0	0	0	0
DBF_3	0	**1**	0	0	**1**	0	**1**	**1**
DBF_4	0	0	0	0	0	0	0	0
DBF_5	0	0	**1**	0	0	0	0	0
DBF_6	0	0	0	0	0	0	0	0
DBF_7	**1**	0	**1**	0	0	0	0	0
DBF_8	0	0	**1**	0	0	0	0	0

d)

Figure 4.13 Network structure for selected periods (bold numbers represent collaborations)

The results on the network dynamics of our artificial biotechnology industries are, at first glance, difficult to compare with data from the real world. In Figure 4.14 we find for a single period (1998) a small selection of collaborations between LDFs and DBFs. However, graph theory (see, for example, Burt 1980) offers some measures to compare different networks from a structural perspective.[3] These measures describe, for example, the adjacency, the reachability and the connectivity of a network as well as the centrality of single actors (see, for example, Freeman 1979). By comparing these figures we will get some first insights as to whether we have caught the basic mechanisms of networking in our industries, or where we will have to modify specific components of our model in order to improve our understanding.

LDF / DBF	AHP	Bayer	Boeh Ingel	Dupont Merck	Eli Lily	Glaxo Wellc.	Höchst	Roche	Merck &Co.	Novar-tis	Pfizer	SKB	Warn. Lamb.	Zen eca
Affymax	2					1	1		1	2				
Affymetrix	1							2			1			
ArQule	2							1						
British Biotech	1					2				1	1	2		
Celltech			1						2					2
Chiron							1	1		1			1	
CoCensys	1									1			1	
Human Genome Sci.								1				3		
Incyte Pharma		1			1		1			1	1	1		1
Millennium Bio Therap.	1				2			1						
Neurogen	1										3			
Onyx		1			1								2	
Repligen					1	1			2	1	2			
Scios				1	1		1	1			1			
Sequana Therap.			1			1		1					1	
SIBIA					1	1				1				
Xenova													2	

Figure 4.14 Collaborations in the biotechnology-based industries

In the following we have applied three concepts, in particular the average distance, a network centralisation index and the degree of centrality in order to compare time series of artificial networks with real networks. The computations are done with Ucinet (Borgatti, Everett and Freeman 1999), a software tool designed for network analysis.

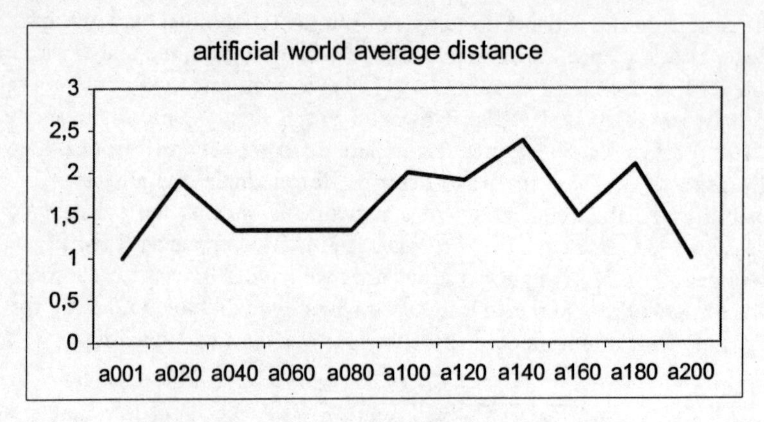

Figure 4.15a Average distance in the artificial world

Figure 4.15 shows the development of the average distance, a measure for the average shortest path between two nodes for our artificial network as well as for our empirical database. This measure can be interpreted as an indicator for the diffusion of information in a network.

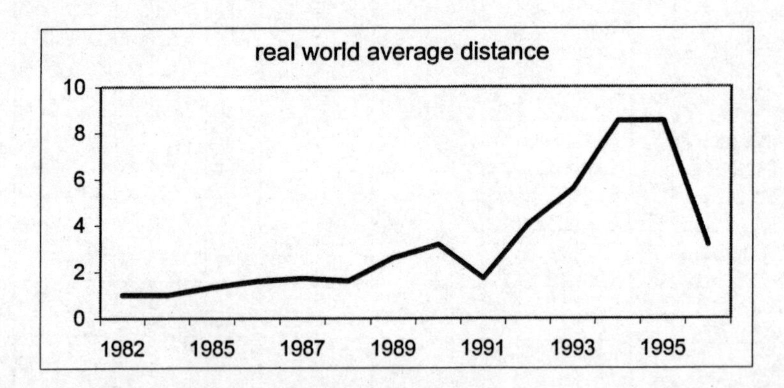

Figure 4.15b Average distance in the real world

First it is obvious that the scale of this measure is significantly larger for the real world compared to the simulated world. However, this measure is an absolute one and depends also on the size of a network. Therefore, the difference in scale can mainly be traced back to the difference in network sizes. Nevertheless, both figures show a structural similarity in a sequence of peaks which indicate a qualitative change in the network structure. Whereas these peaks grow in magnitude in the real world, their artificial counterpart stays

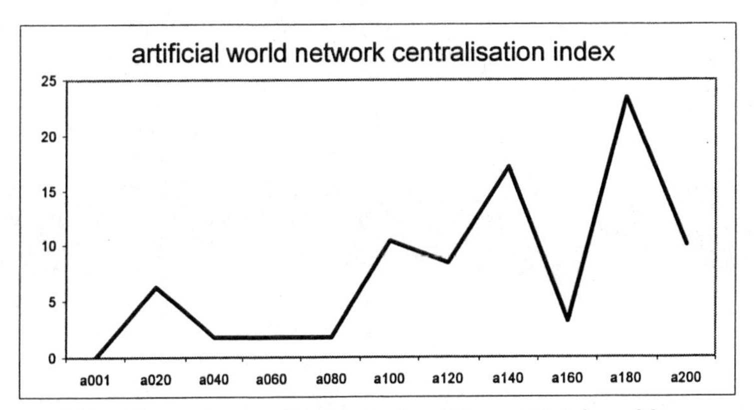

Figure 4.16a Network centralisation index of the artificial world

almost on the same level and also the second peak is unimodal in the real world compared to the bimodal peak in the artificial world.

To rule out the influence of network size, index-oriented measures exist. In Figure 4.16 we apply the network centralisation index which can be inter-

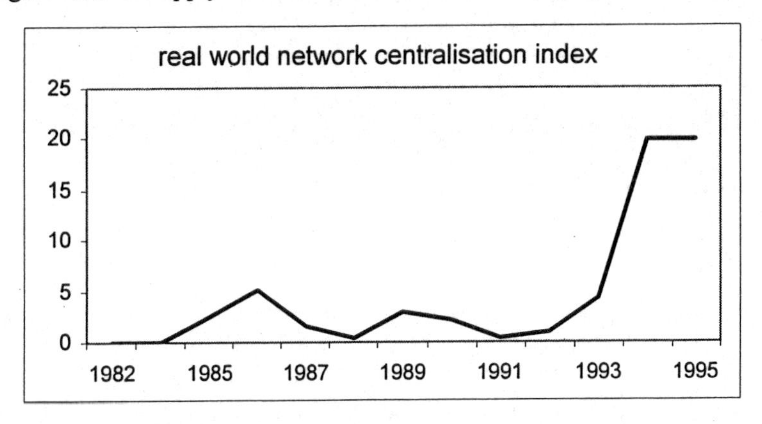

Figure 4.16b Network centralisation index of the real world

preted as a measure for the influence of core actors in a network. Again we find a sequence of peaks for both worlds which now are higher in magnitude in the artificial world. This difference in the impact can still be traced back to the different sample size: in the artificial world we consider for the moment only four LDFs which very likely play the role of core actors. Accordingly, their relative impact in a population of 12 firms is likewise higher compared to a real world firm population of almost 1,000 firms.

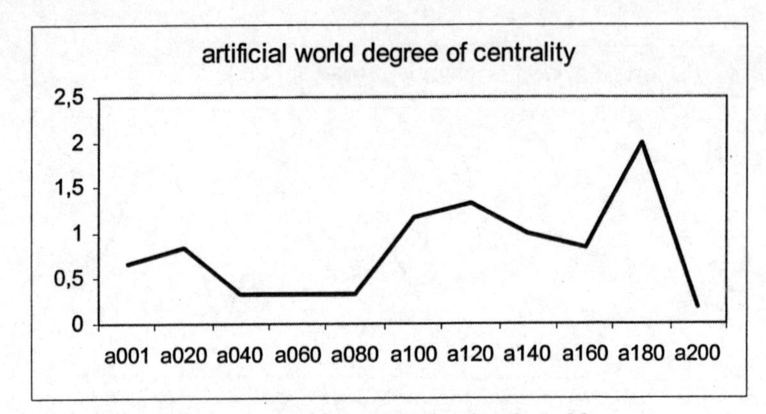

Figure 4.17a Degree of centrality in the artificial world

Finally we measured and calculated the degree centrality for both of our worlds shown in Figure 4.17. The degree centrality measures the asymmetry in the roles played by various actors in a network.

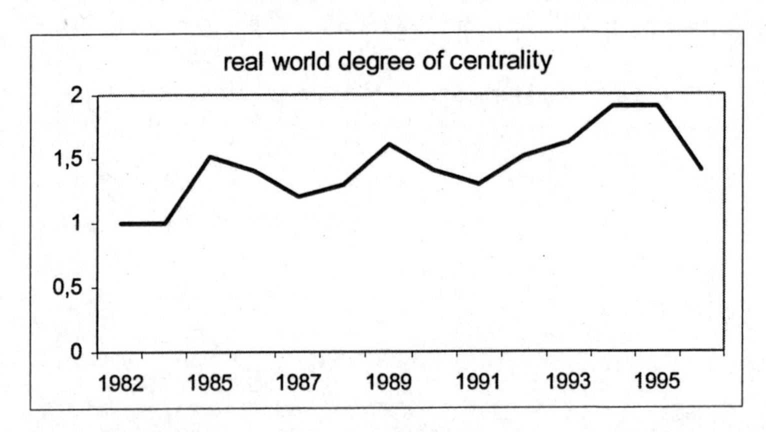

Figure 4.17b Degree of centrality in the real world

Also for this aspect of network activity we find a broad correspondence of our artificial and real worlds. The sequence of three peaks can be interpreted as a consequence of the changing role of DBFs in the networking processes. The first peak is caused by DBFs playing the role of translators supporting the LDFs in their efforts to overcome the gap between their dominant knowledge orientation and the upcoming new knowledge base in biotechnology. The second peak has to be characterised as an intermediate phase, with some DBFs that have already become vertically integrated producers and LDFs still mainly concerned with building up competencies in the new field. The third wave in networking, then, is caused by a tremendous growth in the technological opportunity space, where networking is considered to be a strategy to cope with the speed and complexity of technological development. In this phase, DBFs play the role of explorers allowing the large and established firms to explore a wider range of technological approaches within biotechnology.

CONCLUSIONS

This chapter provides a simulation analysis of the evolution of innovation networks in biotechnology-based industries. Since this is an applied simulation exercise, great emphasis is placed on the characteristic features of this industry. Obviously the implementation of the model in the sense of a history friendly model is not an easy endeavour. The first step therefore was to analyse a prototypical case which allows detection of the numerous mechanisms and interactions.

In a second step the results of the simulations are compared to developments in the real world by applying concepts of graph theory which provide us with some measurements of the overall network dynamics. Although there are still some significant differences between the artificial evolution of network structures and the real world networks, the results look promising as they are able to reproduce at least qualitatively some developments which are observed in reality. The next steps have to be to balance the different mechanisms and to find relative weights in accordance to their specific impacts. Once such weights are attributed, different scenarios can be analysed, showing the influence of different environments as well as of policy measures aiming at the establishment of these new biotechnology-based industries.

One final remark with respect to the stochastic influence of innovation processes on results seems to be necessary. By repeating the simulation experiment several times the Poisson-distributed random number, responsible for the innovation event, leads to varying relationships between the firms in our sample. However, although collaboration partners change, the overall network dynamics do not depend on stochastic influences, but remain rather stable over a large number of simulation experiments performed in a Monte-Carlo-method fashion.

To summarise: in our research we started from the empirical literature and from the existing case studies on biotechnology-based sectors and developed a formal representation of innovation networks that, while abstract, matched a number of the observed features of innovation in these sectors. Working through this analytical exercise has significantly sharpened our theoretical understanding of the key factors behind the development of networking in the biotechnology-based sectors and contributed to a more general understanding of innovation networks in other sectors.

NOTES

1. See, for example, Acharya (1999, 15ff).
2. In this respect, a major methodological advantage of simulation studies shows up in the construction of the innovation processes. Whereas in traditional optimisation models there is no difference between the modeller and the modelled agents, simulation analysis allows programming random numbers as their statistical distribution is unknown to the agents in the model, see Pyka (1999, 189ff).
3. An interesting application of graph theory on biotechnology innovation networks is in Pammolli and Riccaboni (1999).

REFERENCES

Acharya, R. (1999), *The Emergence and Growth of Biotechnology, Experiences in Industrialised and Developing Countries*, Cheltenham, UK and Northampton, MA, USA: Edward Elgar.

Borgatti, S.P., M.G. Everett and L.C. Freeman (1999), *Ucinet 5 for Windows: Software for Social Network Analysis*, Natick: Analytic Technologies.

Burt, R.S. (1980), 'Models of network structure', *Annual Review of Sociology*, **6**, 79–141.

Cantner, U. and A. Pyka (1998), 'Absorbing technological spillovers: Simulations in an evolutionary framework', *Industrial and Corporate Change*, **7**, 369–97.

Cohen, W.M. and D.A. Levinthal (1989), 'Innovation and learning: The two faces of R&D', *The Economic Journal*, **99**, 569–96.

European Commission (1997), *Second European Report on S&T Indicators*, Brussels.

Freeman, L.C. (1979), 'Centrality in social networks, conceptual clarification', *Social Networks*, **1**, 215–39.

Gibbons, M., C. Limoges, H. Nowotny, S. Schwartzman, P. Scott and M. Trow (1994), *The New Production of Knowledge: The Dynamics of Science and Research in Contemporary Societies*, London: Sage.

Grabowski, H. and J. Vernon (1994), 'Innovation and structural change in pharmaceuticals and biotechnology', *Industrial and Corporate Change*, **3**, 435–49.

Kuenne, R.E. (1992), *The Economics of Oligopolistic Competition*, Cambridge, MA: Blackwell Publishers.

Malerba, F., R.R. Nelson, L. Orsenigo and S.G. Winter (1999), 'History friendly models of industry evolution: The computer industry', *Industrial and Corporate Change*, **8**, 3–40.

Nelson, R.R. and S.G. Winter (1982), *An Evolutionary Theory of Economic Change*, Cambridge, MA: Cambridge University Press.

Pammolli, F. and M. Riccaboni (1999), 'Technological change and network dynamics: The case of the bio-pharmaceutical industry', Paper presented at the *European Meeting on Applied Evolutionary Economics*, Grenoble, 7–9 June 1999.

Prahalad, C.K. and G. Hamel (1990), 'The core competencies of the corporation', *Harvard Business Review*, **90**, 79–91.

Pyka, A. (1999), *Der kollektive Innovationsprozeß – Eine theoretische Analyse absorptiver Fähigkeiten und informeller Netzwerke*, Berlin: Duncker & Humblot.

Saviotti, P. (1998), 'Industrial structure and the dynamics of knowledge generation in biotechnology', in J. Senker (ed.), *Biotechnology and Competitive Advantage: Europe's Firms and the US Challenge*, Cheltenham and Northampton, MA: Edward Elgar, pp. 19–43.

Staropoli, C. (1998), 'Co-operation in R&D in the pharmaceutical industry', *Technovation*, **18**, 13–24.

Teece, D. (1986), 'Profiting from technological innovation', *Research Policy*, **15**, 285–305.

Tushman, M.L. and P. Anderson (1986), 'Technological discontinuities and organisational environments', *Administrative Science Quarterly*, **31**, 43.

5. The Role of Knowledge-Intensive Business Services (KIBS) in e-Commerce

Paul Windrum

INTRODUCTION

The focus of the research in this chapter is placed on market-driven processes that construct and co-ordinate the large innovation networks found in e-commerce. The potential importance of e-commerce makes the study particularly interesting. Like technologies such as electricity and the automobile before it, the socio-economic implications of the internet are immense. This is because it is an underpinning technology the development and diffusion of which affects the production, distribution and consumption of many goods and services. In addition, it facilitates the development of new forms of communication, organisational practices, working patterns, and social intercourse and differentiation.

The case studies reported in this chapter examine in detail the activities of a particular type of knowledge-intensive business services, provider of the professional web authoring company. Such companies are active in business-to-business services revolving around the design and implementation of an integrated internet-based platform for conducting e-commerce. The case studies cover five micro networks, each of which comprises a web authoring company, a business client, and an independent contractor. Three of these micro networks are located in the UK and two are located in the Netherlands.

The section that defines knowledge-intensive business services (KIBS) starts to open up, in a generic manner, the discussion of the role played by KIBS in innovation networks. The next section maps out the e-commerce innovation network in which web authoring companies operate. Here the representations of the web authoring companies contacted by the study are used to map out the relationships found in the e-commerce network to which they belong. Within this mapping, the key supply and demand side actors are described, and the key factors that appear to be influencing the development and diffusion of e-commerce systems are discussed. This paves the way for an analysis in the last section of the key findings of the empirical research.

The key findings of the empirical research are clustered within three broad domains of inquiry: the nature of competition in e-commerce networks, the role of standards in e-commerce, and the role played by the web authoring companies in e-commerce innovation networks. These key research findings are, first, distributed knowledge production in e-commerce that involves multiple actors on both the supply and demand side of the market. Second, the interactions between, and knowledge flows amongst, these actors are complicated and non-linear. Third, production of the hardware and software elements that make up e-commerce is disseminated across many firms and no single producer can entirely shape, by its own volition, the evolutionary trajectory of the technology. Fourth, the value of each element making up the technology is interrelated. This has important implications for both producers' innovation strategies and for users' evaluations of the return to adoption. Fifth, the complexity of the e-commerce innovation network is such that it requires integration and mediation. Two key types of system integrator are identified by the research: human actors (for example, web authoring companies), and non-human actors (for example, standards).

KNOWLEDGE-INTENSIVE BUSINESS SERVICES

Knowledge-intensive business services (KIBS) firms are private sector organisations that rely on professional knowledge or expertise relating to a specific technical or functional domain. KIBS firms can be primary sources of information and knowledge (through reports, training, consultancy) or else their services form key intermediate inputs in the products or production processes of other businesses (for example, communication and computer services). As well as being users of new technology, some KIBS are carriers of new technology (for example, consultancies and training services), while others are themselves integral producers of new technologies – notably in computer, software, telecoms and telematics services. Indeed, there appears to be a positive association between KIBS and new technologies. New technologies have spawned new service industries that have in turn played a major role in developing these technologies through laboratory, design and engineering activities. In addition to information and communication technologies such as the internet, notable examples include those KIBS services connected with biotechnology, new materials and environmental technologies.

Antonelli (1988) highlights the link between new Information and Communication Technologies (ICTs) and business-to-business KIBS. He argues that the post-war organisational structure of vertically-integrated knowledge generation, founded on an R&D lab, is being replaced by an information exchange market based on real-time, on-line interactions between customers and knowledge producers. New ICTs facilitate a change in the nature of information – its divisibility, processing and communication – and the accessibility and tradability of information. This opens up opportunities for knowledge-intensive business service firms who, Antonelli adds, tend to be the chief advocates and supporters of this emerging information market. Through ICTs, KIBS companies interface between a client firm's tacit knowledge base and the wider knowledge base of the economy, improving connectivity and receptivity between the nodes of innovation networks. KIBS firms thus acquire special significance as agents who transfer experience and technologies within, and across, innovation networks.

An important feature that distinguishes KIBS from manufacturing firms is the type of 'product' they supply and, following this, the role which they play in regional and national innovation systems. Whereas manufactured products and processes contain a high degree of codified knowledge (they are a 'commodification' of knowledge), KIBS products contain a high degree of tacit ('intangible') knowledge. As well as being vehicles of knowledge transfer, KIBS firms are engaged in the co-production of *new* knowledge and material artefacts with their business clients. This interactive problem-solving is the 'product' that clients wish to purchase. Given the importance of this interaction term, we should not overlook the factors that facilitate successful KIBS-client interaction. The quality of the provider-client interaction depends on the competences of the client as well as the KIBS supplier. In addition to ICT proficiency and other technical competences, clients must be open to new learning contexts. Strambach (1997) suggests that knowledge diffusion is facilitated by flexible, decentralised organisation structures and good quality human resource management. Cohen and Levinthal (1990) point out that a firm's absorptive capacity – its ability to assimilate new information – is closely related to its organisational routines, and the diversity (i.e. the level and distribution) of expertise within an organisation. Meanwhile Ciborra (1993) emphasises the match between new knowledge and the practices, beliefs, values, routines and cultures that lie at the heart of the organisation.

MAPPING THE INNOVATION NETWORK IN e-COMMERCE

It is instructive to open up the discussion of innovation networks in e-commerce by identifying the key participants who, individually and collectively, are shaping its direction. Notable amongst these are users (businesses and households connected to the internet and interacting with it), communication lines and communications equipment providers, intermediaries (the suppliers of on-line information or access providers), hardware manufacturers, software authors and manufacturers (for example, browsers, site development tools, specific applications, smart agents, search engines and others), content producers and providers (for example, media companies), suppliers of financial capital, and public sector institutions. A sketch of an e-commerce innovation network, as seen from the perspective of the web companies interviewed by this study, is given in Figure 5.1 below.

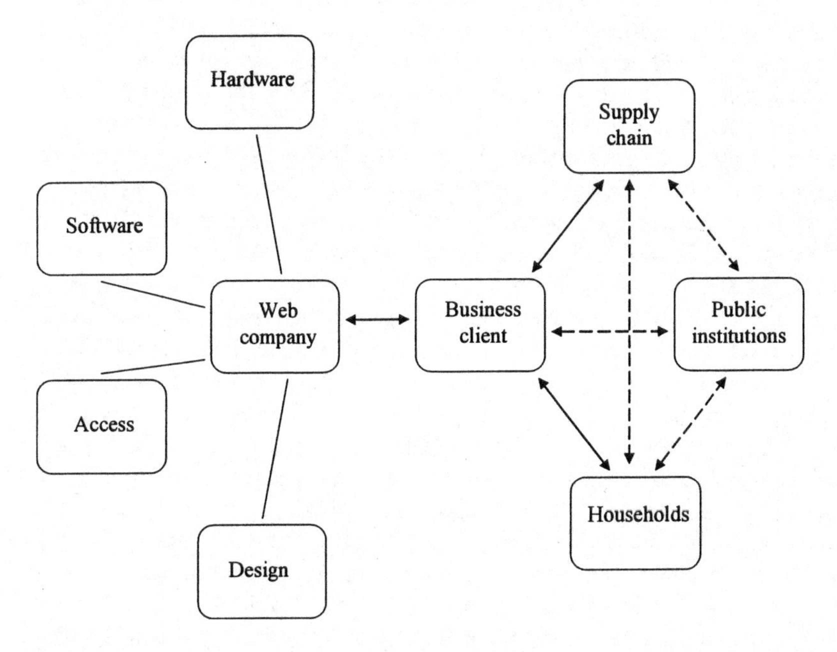

Figure 5.1 A web company's view of an e-commerce innovation network

A web authoring company is a systems integrator who brings together hardware, software, access provision and web design skills to deliver bespoke website solutions for business clients. The look, feel and functionality of the website is co-produced by the web company and its client through a process of negotiation. According to the interviewees contacted by the study, the key competence for successful web authoring companies is that of problem solving. This involves an array of skills and knowledge such that the provision of a tailored solution may require a combination of in-house and externally held skills sets to be employed. Outsourcing is common when highly specialised skills are required. Commonly cited examples of outsourcing in our case study involved the use of freelance designers possessing specialist graphics and database skills, and specialist intranet/extranet contractors. One consequence of the shift towards co-production is the need for web authoring companies to develop project management skills. Indeed, one company interviewed by the study went so far as to view itself primarily as a systems integrator that configures packages of network components to suit the particular needs of its clients.

The supply side of the market is notable for both its variety and its innovative content, with companies moving across multiple product domains over time. However, it is possible to divide the supply side of the web market into three broad categories. The first category encompasses software producers. These include software houses offering both bespoke and standardised software packages. To date the most prominent have been those offering graphical browser interfaces that facilitate access to the Web and third-party providers, such as Macromedia and Progressive Networks Inc., who offer multimedia and other plug-ins for these browsers. The second category encompasses internet service providers. Following the 'privatisation' of the NSFNET on 30 April 1995, the internet became an interconnected mesh of ISPs. An ISP is a company that connects members of the general public to the internet. ISPs have been differentiating themselves into distinct market niches over the last couple of years. 'Backbone ISPs' specialise in BFRS (big fast routers and switches) and high-speed long-haul circuits. 'Dial-up ISPs' specialise in many points of presence (POPs) which accept local dial-ins from clients using modems. 'Back-end ISPs' specialise in web hosting and carrying frequently accessed information to server caches nearer large populations of users. 'Front-end ISPs' specialise in high-performance access and data caching for local user populations. An important retail market has developed with many new entrants from existing industrial sectors – notably telecoms and well-known brand names – moving into the area. The web authoring companies interviewed in this study are also access providers, offering specialist services to corporate business clients.

The third category making up the supply side of the web market is hardware manufacturers. A host of computer hardware, networking companies and cable operators, ranging from Cisco to Hewlett Packard to Cable & Wireless, are contributing to the infrastructure of the internet through provision of routers, servers and fibreoptic links. One does not immediately think of computer hardware manufacturers as actively contributing to the activities of the web authoring companies. Yet hardware manufacturers are keen to monitor changing patterns of demand and have contacts with a number of the web authoring companies contacted by the study. In addition to long lead-times, demand is highly unpredictable and rapidly changing in unanticipated directions. As well as keeping a close eye on market developments, hardware manufacturers are important sources of innovation in their own right.

Turning to the demand side, the diffusion of e-commerce across businesses depends on its potential to improve their competitive performance. Thus far there are few clear exemplars that indicate clear benefits in terms of productivity, costs, sales and profits. This may in part reflect the radical and demanding nature of the e-commerce model for businesses where e-commerce is the integration of the internet and related ICTs into the business organisation. This has two facets. One is the restructuring of the supply chain so that production and delivery become a seamless process. The other is the creation of new business models based on open systems of communication between customers, suppliers and partners. Where the integration of the supply chain provides increased efficiency and significant cost advantages through waste minimisation, the development of new products and services are facilitated by new ways of conducting business based on internet working between organisations and individuals. The importance of the e-commerce website lies

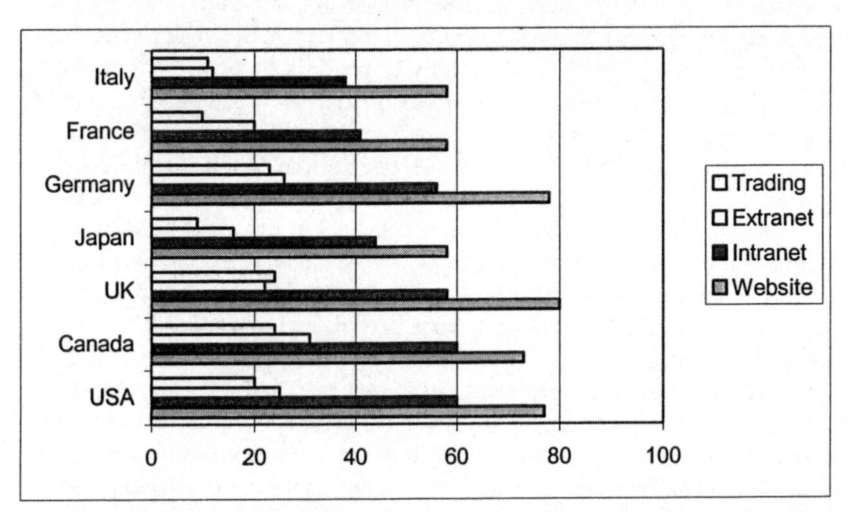

Figure 5.2 G7 countries comparison of the proportion of businesses (weighted by business size) possessing a website, intranet, extranet, and trading online in per cent

(source: *Spectrum/DTI*, 2001)

in its being the hub of an integrated communication network that links supply chains with households. This is achieved by bringing together internet, intranet and extranet information networks.

Data collected by Spectrum/DTI (2001) on businesses across the G7 countries suggests that the majority of firms now have a company website and that the use of intranets is now widespread (see Figure 5.2). Indeed, the lead enjoyed by firms in the USA two years ago has now largely disappeared with respect to website and intranet ownership. This is supported by the findings reported in our case study interviews, which suggest that the web has rapidly diffused within the UK and Dutch business communities over the last three years. Our respondents observed that this is particularly true for large and medium-sized businesses although small-sized and micro businesses still lag behind in their adoption rates, in part due to the initial set-up costs (in time and human resources as well as initial financial costs) and is in part due to the perceived returns on investment. This is supported by the findings of the Spectrum/DTI data that observes a lag between adoption by micro and small businesses on the one hand, and medium and large businesses on the other.

Yet web presence and intranets alone do not constitute e-commerce. Notably, Figure 5.2 highlights the low proportion of businesses that currently provide customer extranets in each of the G7 countries (Canada 31 per cent, Germany 26 per cent, USA 25 per cent, UK 22 per cent, France 20 per cent, Japan 16 per cent, Italy 12 per cent) and the low proportion of businesses actually trading over the internet (Canada 24 per cent, UK 24 per cent, Germany 23 per cent, USA 20 per cent, Italy 11 per cent, France 10 per cent, Japan 9 per cent). Relatively few businesses are actively exploiting the internet as a commercial business tool. This was supported by the case study interviews conducted in this study. As one interviewee put it, 'The business community is not seriously surfing the net. They will inspect their competitors' sites to see if they have a better webpage, but relatively few are conducting real business-to-business activity, such as looking for new suppliers.' In order to understand this we need to consider the degree of complexity associated with establishing an e-commerce website and other factors that have been driving demand.

One can identify three types of websites, each with an increasing degree of technical complexity and sophistication in the communication structures that can be handled. The first and simplest form of company website is the 'flat ad website'. This consists of a one or two page document providing background information on the company and its contact address and telephone number. In the past a flat ad website did not always have an e-mail link, although this is now a standard feature. A far more sophisticated web presence is provided by the second type of website, the 'brochure website'. This provides details of product specifications, price, and availability, and may make use of graphical illustrations of products much in the manner of a mail order catalogue. A brochure website also opens up the possibility of on-line ordering and payment. On-line selling provides a relatively low cost means of reaching a large number of customers. Additionally, the geographical reach of a website is far greater than that of a traditional retail outlet. If a digital commodity is being sold then multiple copies can be distributed across the internet at zero marginal cost. If the commodity cannot be digitised then a means of physical delivery, for example, overnight carriers, needs to be organised. The introduction of on-line selling invariably requires large-scale changes in information systems supporting outbound logistics – including stock and inventory control, warehouse management, and delivery planning and control – and marketing and sales. This often requires the introduction of new complementary ICT technologies, new competences and business practices, and a degree of organisational restructuring.

The third type of website is the fully-fledged 'e-commerce website'. This brings together the internet and other ICTs to support a new form of business model; one that restructures the supply chain so that production and delivery become a seamless process which can be monitored from beginning to end and which enables reconfiguration of the business-customer interface with the aim of constructing an open mode of communication between customers, suppliers and partners along the supply chain. The advantages of this new model are twofold. First, the integration of the supply chain provides increased efficiency and affords significant cost advantages through waste minimisation. Second, new products and services development is facilitated by new ways of conducting business based on the internet working of organisations and individuals. Thus a high degree of organisational restructuring is required for businesses making the transition to e-commerce. This restructuring affects all of the primary activities of the organisation's value chain: inbound logistics, operations/manufacturing, outbound logistics, marketing and sales, and after sales support. The wesbite itself becomes an interactive platform, using technologies such as dynamic databases and video conferencing, to host multilateral communications between clients and providers on product design and a range of other areas. Extranets enable a business to share part of its information or operations with suppliers, vendors, partners, customers, or other businesses. These can be used to check raw material requirements, investigate stock availability, and check the progress of an order. For many businesses this represents an important step in the adoption of new working practices with partners. Internally, new ways of working are also being explored by offering employees remote access. Marketing and sales are further transformed by the development of after sales support underpinned by database records of customers and telephone call centres. Finally, e-commerce requires the introduction of new financial management practices and support systems due to the high degree of complexity, and the need for continuous information collection and monitoring in order to optimise cash flow, in real time.

Returning to Figure 5.2 and to the comments of interviewees contacted by this study, the proportion of businesses possessing some form of website, intranet and extranet, and/or engaged in on-line trading suggests that the majority currently have either a flat ad or a brochure website. Few businesses are as yet operating a fully-fledged e-commerce website. Difficulty of implementation and uncertainty with regards to profitability have been key factors inhibiting the development and diffusion of e-commerce websites. By contrast, basic flat ad and brochure websites are relatively easy to set up. As noted previously, the diffusion of these types of websites has been quite rapid. According to the respondents contacted by this study, this has been fuelled by expectations both within the industries in which the client firms operate and the wider milieu, fuelled by a high level of media attention. As one client company put it, this gave 'A sense that something was going to happen. Rivals were getting on board and had some kind of web presence – however basic – or it was likely they would soon be getting on board. So we needed some basic level of web presence in case we were left behind.' This is an almost classic description of a 'bandwagon effect' with individuals demanding more of a commodity because all other individuals in the market demand more of the commodity (Leibenstein 1950).

RESEARCH FINDINGS

The Nature of Competition in e-Commerce Networks

The various supply-side categories highlight the extent to which innovation is distributed across a multiplicity of contributors. It is in part driven by the activities of software producers, in part by hardware providers and in part by solutions providers who, working together with their clients, explore imaginative new areas of commercial application. Indeed interrelatedness of these innovation networks, in both the production of e-commerce systems and in their consumption, is a particularly striking feature of e-commerce. The systems perspective of technological change and innovation put forward by Rosenberg some 20 years ago appears particularly apt in this context. Rosenberg observed that epochs in economic history are marked by the emergence of clusters of new, interrelated technologies rather than single technologies. 'The importance of these complementarities suggests that it may be fruitful to think . . . of these major clusterings of innovations from a systems perspective' (Rosenberg 1982, p. 59). Technological competition occurs at two levels. At the inter-systems level, producers are seeking to develop new e-commerce systems that will replace older, established systems such as electronic data interchange (EDI) systems for handling business-to-business transactions.[1] At the intra-system level, producers compete for control of key elements that make up e-commerce. Perhaps the most publicised example of this has been the competition between Microsoft and Netscape for commercial control of the web browser (Windrum 2001).

The industrial organisation and inter-firm relationships that exist within a system technology are very different to the orthodox textbook representation. First, the various elements of the system are created by different companies with very different competences and areas of expertise. No single producer can entirely shape, by its own volition, the evolutionary trajectory of the technology. Second, the value of each system element depends not only on its own quality and performance characteristics but also on how well it works in tandem with other system elements, and on the overall performance of the system. A mutual value relationship therefore exists between firms producing a range of interrelated system elements. This mutual value relationship has important implications for firms' innovation strategies: for example, the realised returns to the R&D of an innovating firm are clearly dependent on the R&D activities of other related firms, which may be beneficial or detrimental in their effect. This places on the agenda a strategic trade-off between co-operative and competitive behaviour between firms. Third, an opportunity exists for intermediaries, such as web companies and other solutions providers, to offer specialised services that combine hardware, interfaces, software and web design to meet the particular needs of their business clients.

Interdependence of value has important implications for the end-user. If the perceived value of an individual component depends on how well it works in conjunction with other hardware and/or software components, then changes in the specification of one component can have a marked affect on the user's perceived value of the other components making up the technology. Users therefore take an active interest in who produces the individual system components and their compatibility with the elements produced by other firms. Users may wish to have control over the matching of different elements and may have distinct preferences in components derived from a specific supplier. In other words, brand loyalty can become an important factor. For example, there are those who strongly support the development of Linux as an alternative operating system to Windows. On the other hand, there are those who prefer to purchase Microsoft products. The implications of this interdependence of value is distinct from that described by Katz and Shapiro (1985). They point out that the perceived value of a technology depends on the number of users who also adopt the same technology. For example, the value of a telephone system to a user depends on how many other users can be communicated with. Hence the greater the number of users adopting a particular technological variant, the more attractive it becomes to potential adopters, leading to a bandwagon effect. However, in the case described by Katz and Shapiro, the end-user is unconcerned as to how the product is produced – whether by a vertically integrated firm or by a set of distinct but interdependent producers. It is likely that both of the demand-side phenomena described will be present in markets for network technologies.

The Role of Standards in e-Commerce

The qualitative research conducted by this study highlighted the importance of standards for innovation in the e-commerce network. Collective innovation is not restricted to formal inter-firm R&D collaboration. Indeed firms seek to co-ordinate the supply and design of related system components through a number of procedures besides formal collaboration agreements. One avenue highlighted by the study is the setting of compatibility standards – physical interfaces and communication protocols – that ensure interoperability between products when connected together. Process standards governing the ways in which component products are themselves produced (for example, software languages and tools) also play an important role. Responsibility for the formal codification and monitoring of these standards is given over to formal institutions. Some, such as the International Standards Organisation (ISO) and the British Standards Institute (BSI) are government-sponsored institutions while others, such as the World-Wide Web Consortium (3WC) and the Internet Engineering Task Force (IETF), are industry-run consortia. The formal technical standards set by and monitored by these institutional bodies are known as *de jure* standards. In addition to *de jure* standards, there exist market-driven *de facto* standards. These are standard products, often the winners of market competitions between rival alternative designs, whose product characteristics become the industry norm. The WinTel personal computer is an example of a *de facto* standard that is familiar to almost everyone. Its very name denotes the particular combination of software and hardware products – Microsoft's operating system and Intel chip sets – that together define the product. By controlling the direction of innovation in the operating system and the chip set over successive vintages, the two companies maintain their control over the standard.

In addition to *de jure* and *de facto* standards, the study has highlighted the importance of two other types of standard. First, the informal design conventions – the norms and practices of design in Western cultures – that are embodied in the design conventions found on websites. Second, the ergonomic 'look and feel' qualities of user interfaces. These are aspects of aesthetic design which determine the appearance of a product, its access conditions, and its ease (or difficulty) of use – aspects not dictated by design norms and practices. Rather they are arrived at over time through trial and error as designers experiment with new designs and govern feedback from users. Hence various categories of 'standards' were highlighted in the empirical study. Capturing these various categories within a single definition requires a broad definition. For this reason it is useful to follow the definition of Swann and Windrum (1998): a standard as a set of common features, belonging to a group of products and services which benefit users and facilitate compatibility between those components.

The interviewees contacted by the study suggest that standards are important for three reasons. First, they are essential for the integration and development of a complex technology like e-commerce. Second, they promote growth and efficiency. Third, they are a means through which individual firms seek to influence the wider innovation process. The establishment of standards covering product characteristics (what a product is), minimum attributes (what it does), compatibility (what else it can connect with) and ergonomics (how a user interfaces with it) enables each product to be black boxed. Provided an alternative product design meets the same set of standards, the rest of the system will function, regardless of what is actually inside that product. This process of black boxing facilitates modularity and is essential if a technology system is to function. Modularity of system components means innovative effort can be effectively focused on bottlenecks – changes to, or the replacement of, one product can be made without having to simultaneously change all the other system products with which it interacts.

The early identification of interface standards has played a key role in promoting the growth and development of the e-commerce network. As previously noted, it is unlikely that one company, or even a consortium of companies, will be able to supply all the required component elements that make up e-commerce. Instead a large number of companies, some of them small, one-product entrants, are specialising and supplying particular products. At the industry level it has been important to identify a number of the most promising directions for technological development from the myriad of possible options. According to economic theory, the earlier that this is done the greater the chance of achieving a common standard. Delays can result in difficulties in securing a common standard and several incompatible alternatives may remain. Although it is not necessarily disastrous if two or three incompatible standards share the market, such an outcome will reduce the positive externalities enjoyed by all users. A split market may also make it hard for the smallest entrants to survive because they, above all, find it difficult to meet the fixed costs of translating their products to each competing standard.

Innovation within e-commerce is associated with several layers of standards. Maximising the overall performance of a technology system such as e-commerce requires, amongst other things, the maximisation of their interaction. Interface standards are thus as important as the *de jure* standards defined by standards institutions and *de facto* standards that emerge through market competition. Interface standards can take two forms. One is the interface between physical hardware or software components, the other is the 'look and feel' design interface which governs user-product interaction. If the latter is common across a number of products, then a user will be better placed to make use of the other components within the system technology. The creation of new commercial markets can therefore be assisted by the establishment of standards bodies that specify codes of behaviour or rules governing the quality of design or hardware/software interfaces.

Finally, if the establishment of a standard influences and shapes the collective innovative behaviour of e-commerce producers, then considerable power is conferred on those firms who are able to set and control standards. Standards have become a key strategic device by which individual firms seek to gain influence over the wider innovation process. Not only does a standard introduce a degree of certainty in the behaviour of others but it makes that behaviour conform to a common framework of reference that is more readily understood by the standard-setter. Returning to the example of the WinTel PC, its very name indicates that the conceptualisation of what this technology is, how it should operate, and its future technological trajectory is effectively set by two firms: Microsoft and Intel.

Challenges Faced by Technology Users

It has been argued by von Hippel (1987) that the user can be an essential source of innovative ideas, even if the implementation of those ideas is carried out by a producer or a solutions provider. However, the case study interviews suggest that innovation in e-commerce has so far been driven by suppliers. Through experience, businesses are becoming more aware of the commercial opportunities afforded by the internet and are increasingly viewing it as a means of securing a competitive advantage. Indeed the web authoring companies and freelance designers interviewed by the study noted that the distinction between first generation business adopters and those businesses with previous experience manifests itself in a number of ways. One obvious distinction is in motivation. In general, new users are simply interested in 'getting on the internet' and so are happy to have some very basic form of web presence. By contrast, businesses with past experience seek to achieve something new or novel with the internet. By their own admission, new businesses clients are often motivated to invest in their first website because they see their competitors making this investment. Internally, top level management are rarely actively engaged in the development of the first generation website. This tends to be placed in the hands of one person, usually drawn from lower or middle management, who then acts as a technology gatekeeper for the business having responsibility for the design and maintenance of the website. If the first generation website is deemed a success then marketing, sales and other key departments begin to take an active interest in its commercial development. This shift becomes particularly noticeable, we were informed, when the website is due for a major redesign. The specification of the second generation website invariably involves more senior management, from a series of departments, who seek to shape the character and content of the website.

When probing the issue of why innovation has been supply driven, the case study respondents highlighted three sets of challenges facing business users: the degree of organisational change required by the transition to e-commerce, the radical restructuring of businesses' communication modes, and high levels of ongoing technological change. Returning to the discussion of flat ad, brochure and e-commerce websites , it is possible to trace a number of stages through which firms are passing as they progress towards e-commerce. Applying Porter's value chain model to the discussion, we see that higher degrees of organisational restructuring are associated with each of the stages to e-commerce. Setting up a flat ad site with an e-mail link is relatively easy because its introduction only affects those internal information systems that are linked to support activities in the value chain. By contrast, on-line selling within a brochure site requires a higher degree of organisational restructuring because its introduction affects outbound logistics and marketing, both of which are primary activities in an organisation's value chain.

A still higher degree of organisational restructuring is required for businesses making the transition to e-commerce itself. As previously noted, e-commerce involves both the restructuring of supply chains and the reconfiguration of the business-customer interface with the aim of constructing a seamless web between customers and suppliers along the supply chain. Hence the introduction of a fully-fledged e-commerce website requires a restructuring of the primary activities of the organisation's value chain: inbound logistics, operations/manufacturing, outbound logistics, marketing and sales, and after sales support. The internet provides a common, interoperable platform for this new business model. A set of additional elements arise in the e-commerce framework. Interactive websites, incorporating technologies such as dynamic databases and videoconferencing, facilitate multilateral communications between client and provider regarding product design. Extranets enable a business to share part of its information or operations with suppliers, vendors, partners, customers, or other businesses. These can be used to check raw material requirements, investigate stock availability, and check the progress of an order. For many businesses this represents an important step in the adoption of new working practices with partners. Internally, new ways of working are also being explored by offering employees remote access. Marketing and sales are further transformed by the development of after sales support underpinned by database records of customers and telephone call centres. Finally, e-commerce requires the introduction of new financial management practices and support systems due to the high degree of complexity, and the need for continuous information collection and monitoring in order to optimise a business's cash flow, in real time.

The business users contacted by the study highlighted the difficulties faced in creating an integrated e-commerce system. As one business user put it:

> Developing a real e-commerce system requires two things. Both involve a radical change in a company's organisational structure, and both can be really difficult to get accepted. First, you need to bring together lots of data processing activities that were previously separate. This is good because you really see where your systems were not integrated. But it also means you have to deal with lots of departmental politics. Department Heads are always suspicious about anything that looks like it'll threaten their autonomy. The second thing is that, once you've brought all these processes together, you then want to expose them to the outside world through the web! That can easily show up the inefficiencies and inconsistencies that were previously hidden from your customers and your suppliers. Not surprisingly, that also meets a lot of resistance . . . your arguments about the benefits of integrated e-commerce systems need to be convincing before starting the whole process. People have to be convinced and brought on board from the outset.

The different types of communications required for e-commerce have been highlighted in the literature (for example, Venkantraman 1994; Gonzalez 1998). The internet is an inherently two-way medium that requires the development of new styles of conversation between an organisation and its clients/suppliers. It is still possible for organisations to use the internet to conduct uni-directional interactions, for example, completing and dispatching forms, sending and receiving messages via e-mail or v-mail). However, what is novel about the internet is the ability to converse with someone while both parties work on an application and see the conversant and transfer documents as the conversation continues. Gonzalez (1998) distinguishes between four types of communication, each requiring a higher degree of organisational sophistication: the publication mode, the asymmetrical mode, the symmetrical mode, and the synchronous virtual environment mode. The publication mode is the traditional uni-directional model of previous ICTs in which the sender formulates a static document that the receiver reads. The asymmetrical mode is a bi-directional, time delayed didactic communication with one participant formulating a written question or statement and the other responding after a time lag. The symmetrical mode differs from the asymmetrical mode in that it has multi-directional communication with numerous feedbacks that takes place in real time and which extends the degree of interactivity. Finally, the synchronous virtual environment mode uses real time, dynamic, multi-dimensional communication to support key business processes.

Unfortunately it appears that businesses are having problems in developing a different mode of communication. As one web author put it, the material that clients ask to be placed on their website

> typically reads like a conference paper, with a speaker standing on a podium addressing an attentive audience which is 'out there' . . . This is not the situation they are in! People are coming to your door. When opening the door you must make a good impression. You don't start throwing a speech at them. Instead you open up conversational bridges with them – you say 'Welcome, come in, how can I help you' . . . to have a meaningful interaction the companies *must* start to provide the necessary information and get beyond the brochure mentality.

In addition to organisational and communication issues, technological issues have inhibited businesses from experimenting with fully-fledged e-commerce systems. Rapid rates of technological change over the last five years mean that investments are high, not only in terms of finance but also in time and human resources. Moreover, returns on investment have been lowered since skills sets and competences require regular updating preventing economies of scale in scale and scope. While technological change has been rapid in e-commerce, some key technological problems are as yet unresolved. Internet security is a particular case in point. An e-commerce system extends from a company's back-end systems and raw material suppliers through to a range of service delivery channels. This means the web of electronic links between the customer ordering from a website through to suppliers' processing order applications must be secure and trustworthy. However, despite repeated promises over the last couple of years, and formal collaboration between a variety of (normally competing) companies such as American Express, Mastercard, Visa, Microsoft and Netscape, a common 'secure electronic transactions' (SETs) system conforming to accepted internet standards has failed to emerge. This has left on-line consumers with little choice but to use credit card numbers or else one of the various electronic cash schemes currently available. Meanwhile, firms are struggling to move beyond the EDI concepts developed over a decade ago. The problem is one of under-standardisation. The development of e-commerce requires a single, universal SET standard. A number of alternative solutions are being offered but none is sufficiently attractive to take on the mantle.

The Roles Played by the Web Authoring Company in e-Commerce Innovation Networks

Our research has brought to light two key roles played by web authoring companies within e-commerce innovation networks. These are, first, the role of KIBS innovator and, second, the role of a system intermediary. Two key research questions addressed by the study were 'To what extent are web authoring companies shaping the commercialisation of e-commerce?' and 'How important is the client-provider interaction in shaping this process?'. A supporting research question was 'What are the core competences of these companies, and are they changing?'. In order to open up this agenda, web authoring companies and their clients were each questioned about 'what' was actually transferred during the creation of a website – physical artefacts, knowledge, training, etc. – and about the nature of their ongoing relations. Here it was essential to pose the same set of questions to both companies and clients in order to take into account their possibly differing interpretations of what was actually transferred and the economic value of that transfer.

As well as being important agents for technological diffusion, the case studies highlight the extent to which web authors are innovators in their own right. It would be easy to misinterpret their role as simply one of assembling a collection of prefabricated parts that have been produced by hardware and software companies. When technological evolution is characterised by mutual interdependencies between different core technologies making up a system, new opportunities can be created through the fusion of what were previously thought of as unrelated knowledge fields. Cross-fertilisation and its importance for technological innovation has been discussed at length by historians of technological change such as Basalla (1988), Mokyr (1990) and Kodama (1996). So, at one level, the contribution of these knowledge-intensive companies lies in their experimenting with novel combinations of system components. At another, it lies in providing clients with a range of services, wrapped up within a website, that are tailor-made to the client's needs.

The importance of incremental learning via trial-and-error experimentation over a series of projects was emphasised by both web companies and freelance designers. Innovative activities are first and foremost driven by the need to solve clients' problems. The website is the end-product of an interactive problem-solving activity that occurs between provider and client. It is as part of this process that web designers take pieces of software/hardware technology and recombine them in novel ways, extending their performance or even creating completely new hybrids that were not considered by their initial developers. In this way combinatorial innovation can lead to distinctly new softwares, practices and designs. Once more this indicates the need to move away from the traditional theory of innovation in which invention is thought to be the preserve of a single producer. Such a model is inappropriate to the study of innovation in e-commerce.

The case studies highlight that, while web design is constrained by technical limitations associated with contemporary software and hardware, these technical factors cannot explain the look and feel of web environments. The most important determinants governing the look and feel of user interfaces are design conventions. These conventions need to be recognised as standards in their own right, for they are a collectively held cluster of ideas and notions regarding what qualifies as 'good design' or 'good design practice'. Sometimes these conventions can seem peculiar to outsiders. For example, why should navigation bars be placed along the left-hand side of a screen and not the right? When pressed on this particular convention, a number of web designers sought to justify it on logical or naturalistic grounds. When pressed further, they tended to resort to engineering 'tried and tested' or 'it's what works best' type arguments. It subsequently became clear that these designers were referring to well-established conventions and traditions in design, some of which may be centuries old. It therefore seems quite 'natural' for these designers to transfer these conventions from traditional media to new media such as the web. These conventions are important harbingers of a deeply structured path dependency that imparts a degree of continuity across successive technologies: from calligraphy to the printing press, to the typewriter, to the PC, and now to the internet.

Successful experiments in new website design diffuse rapidly, with emulation highlighted as a key contributing factor for a high rate of diffusion. Indeed, all the designers contacted by the study freely admitted 'cherry picking' the ideas of others as part of the normal process of developing an initial set of design ideas. As one interviewee put it, designers are 'continually scrutinising each other's work, taking ideas and getting a sense of 'where the next fashion is coming from', what is 'in' and what is 'out'. In the main this tends to involve the incremental development of a set of design ideas, reformulating these to suit a specific purpose. However, a truly novel concept is occasionally introduced by a design group and this is quickly imitated by others in the community. Far from having a detrimental effect on innovation, this culture of emulation appears to have raised the rate of innovation, leading to radical changes in design and the establishment of new design standards for websites. Trends and fashions have been a notable feature of web design over the past five years, with design standards following one another in quick succession. Often initiated by the introduction of a new piece of software technology, such as Frames, Java applets and graphics software, designers have quickly applied these in various different and novel ways. Sometimes, the interviewees noted, designers have been overzealous in their use of new design possibilities, such as moving graphics. However, after a while this explorative process settles down as a consensus emerges on what are 'useful' and 'tasteful' applications. The practices that are deemed to be extreme are generally dropped.

The triumvirate of novelty, recombination and emulation leading to closure on design standards is not unique to the web design community. It is, as Windrum and Birchenhall (1998) argue, a key aspect of population learning and technological innovation in general. What is notable is the speed with which a succession of design fashions has radically transformed website design within such a short space of time and the particularly relaxed attitude taken by web designers towards intellectual property rights (IPR). Indeed this relaxed attitude was also noted by a number of designers' clients, whose own enforcement of IPR protection is much stronger than that of web authoring companies. When probing deeper into this issue with designers, it became evident that an ethos exists which clearly distinguishes between what is 'acceptable practice' amongst fellow designers and what is an 'infringement' of intellectual property. Notably, a strong distinction is drawn between what designers regard as 'creative emulation', on the one hand, and blatant 'copying' on the other. A particularly interesting finding of the study was the extent to which this ethos cuts across the boundaries of web authoring firms. Designers are simultaneously members of a web authoring firm and members of another group – a design community. The design community has its own set of norms, values and practices. Hence, in addition to the norms and practices associated with the need for commercial businesses to protect their intellectual property, designers are influenced in their daily work activities by the norms and values of this 'other' design culture. It would be interesting for future researchers to consider instances where these two cultures clash and how tensions are resolved.

The influence of the UK design community is evidenced in other ways. Notably, the style of UK web design is distinct in character from that found elsewhere in Europe or in the USA. Many design concepts and formats are being transferred from the 'old media' to the internet and it is not by chance that design standards in the UK lean heavily on UK magazine styles. All of the designers (freelance and company designers) contacted by the study have a background in magazines and newspapers, and some, notably the freelance designers, continue to have an involvement in the older media. This helps to explain why the distinctive styles developed by UK magazine designers in the 1980s are now found on UK websites. Taken together, attitudes of web authoring companies to IPR and to design suggest a deeply structured path dependency is present in their innovative behaviour. Where the innovation literature has highlighted the extent to which innovation trajectories are influenced by the early success or failure of companies in a fledgling industry, the case material of this study additionally highlights the extent to which trajectories can be influenced by antecedent technologies or by the personal histories of those engaged in the innovative process. Notably, the study has highlighted how the personal histories and prior training of web designers continue to influence the nature and direction of their innovative activities.

In addition to their role as innovators within e-commerce networks, the study has highlighted a second role played by web authoring companies, that of a market intermediary. In putting together and experimenting with different designs and software/hardware components, web authoring companies spend a significant proportion of time in educating clients on the commercial potential of the internet and on demonstrating pre-existing commercial opportunities. In bridging the needs of users with the latest technical innovations from suppliers, web authoring companies need to be sensitive to user feedback. As users become more experienced, moving from first-generation to second-generation websites, so their feedback and expectations become an increasingly important input into website development. As an intermediary between users and hardware/software producers, problem solving is the key competence of a successful web authoring company. Translating the branding and e-commerce models of client companies can require web authors to draw upon a wide range of technical and design competences. These may be held in-house. Alternatively, some relevant skills may lie outside the company. As competition has intensified, web companies have needed to differentiate themselves to survive and prosper. While there are those who continue to specialise in website design, it is noticeable that a number of the companies contacted by the study are transforming themselves into system integrators. These systems integrators offer clients an entire package of services: from the selection of servers and other hardware, to the installation of network cabling, to the registering of internet names, to the design of an integrated information network that links the public internet with private intranets and customer extranets. Systems integrators such as these play an essential role in facilitating and orchestrating Mode 2 innovation networks in e-commerce.

As a key market intermediary, the selling practices of web authoring companies have an important bearing on market creation. Good selling practice is important for the long-term growth of any industry since the actions of one supplier shapes the subsequent prospects of other suppliers – for good or ill. Both web authors and freelance designers observed that clients with no prior experience of the internet lack the knowledge necessary to assess the relationship between price and quality, and therefore are unable to distinguish between good and bad offers. This leaves them prey to 'cowboy outfits' that are all too willing to exploit customer ignorance. These outfits undercut the competition in order to make short-term private gains, delivering substandard products to the end-user. Since those most at risk tend to be first-time buyers, cowboy selling can create significant negative externalities. Not surprisingly, reputable web companies and freelance designers express a deep resentment towards such sellers. By creating dissatisfied buyers, these outfits are, as one interviewee put it, 'poisoning the well for others'. For this reason, reputable web authoring companies with overloaded order books prefer to recommend other reputable companies, thereby reducing the possibility of business clients encountering bad selling practice.

The implications of bad selling practice for market creation have been examined by Swann and Windrum (1998). The sale of inferior quality products and services is privately profitable for a company that is only concerned with short-term profits and which has no regard for reputation or repeat purchasing. However, the actions of the bad seller retard the diffusion process, thereby damaging the long-term development of the market. Cowboy selling diminishes the returns on the long-term investments made by those businesses that, taking a long-term view of the market, make large sunk cost investments in skills and competences, and in understanding the activities and needs of clients in order to build a reputation for quality. These sunk costs require repeat purchasing in order to secure an economic return. In the worst case scenario bad selling practice can completely drive out good sellers (Gresham's Law). It is therefore puzzling why none of the internet standards bodies that currently exist are concerned with good selling practices.

The absence of a professional body dealing with quality standards accreditation may reflect the difference between private and social returns. While each creditable web authoring company can be aware of the benefits that a standards institute or trade association would bring, and would willingly sign up to such a body, each may be unable to bear the cost of setting up and administering it. The costs of setting up and running a standards body can be significant. For example, the 26 software process standards issued by the Institute of Electrical and Electronic Engineers (IEEE) are estimated to have cost \$6.5 million to develop (Tice 1988). Web authoring companies in the UK are, by and large, small business concerns with limited financial resources. This appears to open the way for a publicly funded standards body in order to bridge the division between social and private returns.

SUMMARY

A major step in realising the objective to address the policy issues raised by Mode 2 knowledge production is the development of a theoretical understanding of (or at least an appreciation of) the emergence of innovation networks and their dynamics. To that end, this empirical case study has sought to identify some of the deep-seated processes and features of the e-commerce innovation network. Five key findings arise from the study. First, distributed knowledge production in e-commerce involves multiple actors on both the supply and demand side of the market. Second, interactions between, and knowledge flows amongst, these actors are complicated and non-linear. Third, production of the hardware and software elements that make up e-commerce is distributed across many firms and no single producer can entirely shape, by its own volition, the evolutionary trajectory of the technology. Fourth, the value of each element making up the technology is interrelated. Fifth, the complexity of this technological system is such that it requires integration and mediation. Two key types of system integrator have been identified by the study: human actors (web authoring companies), and non-human actors (standards).

To begin with, it was observed that distributed knowledge production in innovation networks is associated with technological systems that comprise multiple, interrelated products and services and that e-commerce represents a new means of structuring the production, distribution and consumption of commodities. If the network is to succeed in displacing established technologies that fulfil these functions then it must unpick the cumulative advantages enjoyed by the rival innovation networks who champion those established technologies. The self-organisation of the e-commerce network is highly complex. Indeed, the discussion of hardware, software and content highlighted the extent to which the internet is underpinned by an innovation network that contains a rich diversity of producers, users and public institutions. The internet is, in a true sense, a 'network technology' that has been shaped by the interactions of multiple actors.

The component elements that together comprise the underpinning technology are produced by different companies with very different competences and areas of expertise. Under the conditions of a network technology, no single producer can entirely shape, by its own volition, an evolving technology. Rather, the value of one system component depends on the interrelated components produced by other firms. This mutual value relationship has serious implications for firms' innovation strategies and for their realisation of investment. In terms of innovation strategy, interdependency of value places on the agenda a strategic trade-off between co-operative and competitive behaviour. Furthermore, from the user's perspective, performance not only depends on the quality and performance of a particular component but also on how well it works with other hardware/software components. Changes in the specification of one component can have a marked affect on the user's perceived value of the other components making up the technology.

Perhaps the most novel and important finding of this study is the need for a system integrator in a Mode 2 innovation network. This finding may not be intuitively obvious. Certainly it was not raised by Gibbons et al. (1994). Indeed, in the description of the e-commerce innovation network – with its variety of public and private sector actors and interests, at times acting with remarkable degrees of autonomy and whose direct connections with certain types of actors may be very weak and indirect – it is easy to overlook the issue of system integration. Yet, like any system, the e-commerce innovation network requires integration if it is to function. System integration is essential for the realisation of investment by producers, and of value by users. It is a particularly pressing issue given the interdependence of value in this innovation network.

Two forms of system integrator were identified in the study. One is standards, the other is organisations that take on the role of system integrators. With regards to standards, the study identified the relevance of a range of formal and informal interface standards to the integration and development of the e-commerce innovation network. From an economic standpoint, interface standards are central to any discussion of network externalities. Maximising the externalities associated with a set of system components requires a maximisation of the degree of interconnection between those components. Economies in user learning arise when these user interfaces are shared by a number of products making up the network technology. Furthermore, as noted on various occasions, one-component producers make important contributions to the process of collective innovation within the e-commerce network. The economic viability of these companies depends on the strong network externalities associated with common standards. In the absence of common standards the market would be too fragmented to afford them a profitable niche.

In addition to standards, there are human agencies who act as market intermediaries. One particular type of intermediary, the web authoring company, has been the focus of this particular study. These knowledge-intensive service providers bring together the supply and demand sides of the market, a key interface between the evolving preferences of business clients and the rapidly changing technologies of e-commerce. As the study has highlighted, their innovative activities operate at a number of levels. As well as being an important agency for the diffusion of new technological ideas for their business clients, they are systems integrators who bring together a range of different hardware, interfaces, software and artwork in order to configure bespoke packages tailored to their client's particular needs. In addition, web authoring companies are innovators in their own right. Web authors experiment with novel combinations of system components, extending the performance of new hardware/software and even creating completely new hybrids that were not considered by their initial developers. In this way combinatorial experimentation sometimes leads to distinctly new softwares, practices and designs. Finally, web authors play a key role in shaping both the design standards and ergonomic standards found on websites. The rate of innovation in this domain has been extremely high over the past five years, with novel website designs diffusing rapidly. As the case study highlighted, the relatively relaxed attitude of the web design community towards IPR is notable. Hence a triumvirate of novelty, recombination and emulation has facilitated a very high rate of innovation in website design. Web authoring companies thus provide a good example of the important role played by systems integrators in Mode 2 innovation networks.

NOTE

1. EDI permits the communication of structured messages between customers and suppliers using agreed formats, and allows data to be fed directly into an electronic business process.

REFERENCES

Antonelli, C. (1988), 'Localized technological change, new information technology and the knowledge-based economy: The European evidence', *Journal of Evolutionary Economics*, **8** (2), 177–98.

Basalla, G. (1988), *The Evolution of Technology*, Cambridge: Cambridge University Press.

Ciborra, C.U. (1993), *Teams, Markets, Systems: Business Innovation and Information Technology*, Cambridge: Cambridge University Press.

Cohen, W.M. and D.A. Levinthal (1990), 'Absorptive capacity: A new perspective on learning and innovation', *Administrative Science Quarterly*, **35**, 128–52.

Gibbons, M., C. Limoges, H. Nowotny, S. Schwartzman, P. Scott and M. Trow (1994), *The New Production of Knowledge*, London: Sage.

Gonzalez, J.S. (1998), *The 21st Century Intranet*, London: Prentice-Hall.

Hippel, E. von (1987), 'Cooperation between rivals: Informal know-how trading', *Research Policy*, **16**, 291–302.

Katz, M. and C. Shapiro (1985), 'Network externalities, competition and compatibility', *American Economic Review*, **75**, 424–40.

Kodama, F. (1996), *Emerging Patterns of Innovation: Sources of Japan's Technological Edge*, Cambridge, MA: Harvard Business School Press.

Leibenstein, H. (1950), 'Bandwagon, snob and Veblen effects in the theory of consumer demand', *Quarterly Journal of Economics*, **64**, 183–207.

Mokyr, J. (1990), *The Lever of Riches*, Oxford: Oxford University Press.

Rosenberg, N. (1982), *Inside the Black Box: Technology and Economics*, Cambridge: Cambridge University Press.

Spectrum /DTI (2001), *Business in the Information Age: International Benchmarking Study*, available at <http://www.ukonlineforbusiness.gov.uk/>.

Strambach, S. (1997), *Knowledge-Intensive Services and Innovation in Germany*, Report for TSER Project, University of Stuttgart.

Swann, G.M.P. and P. Windrum (1998), *The Role of Standards for Collective Invention in Network Technologies*, Colline Working Paper, May 1998 (Revised, October 1998).

Tice, G. (1988), 'Are software standards wasted efforts?', *IEEE Software*, **8**, 14–17.

Venkantraman, N. (1994), 'IT-enabled business transformation: From automation to business scope redefinition', *Sloan Management Review*, **35** (2), 73–87.

Windrum, P. (2001), 'Late entrant strategies in technological ecologies: Microsoft's use of standards in the browser wars', *International Studies of Management and Organization*, **31** (1), 87–105.

Windrum, P. and C. Birchenhall (1998), 'Is life cycle theory a special case?: Dominant designs and the emergence of market niches through co-evolutionary learning', *Structural Change and Economic Dynamics*, **9**, 109–34.

6. Innovation Networks and the Transformation of Large Socio-Technical Systems: The Case of Combined Heat and Power Technology

K. Matthias Weber

INTRODUCTION

Rather than looking at the development of an individual innovation network, this case example of the innovation and diffusion of Combined Heat and Power technology (CHP) investigates the co-evolution of technology innovation in networks and of the restructuring of the large-scale infrastructure system in which these networks are embedded; in this case energy supply.

The energy supply systems in most European countries have undergone a process of deregulation and liberalisation. Not least due to the European Union's Single Market policy, principles of competition have been introduced to open up markets and thus to improve economic efficiency. Other policy measures further contribute to the transformation process; for example, environmental regulation and technology policy, with the objective of improving the environmental performance of energy supply.

CHP is widely regarded as a very promising option in environmental terms. It also has several types of application in economic terms (see Appendix 6.1). Although most governments in Europe tend to agree on the desirability of a more widespread uptake of CHP, quite different policy strategies have been pursued in different countries. And, similarly, the role and importance of innovation networks has differed.

This case example is also characterised by the joint consideration of innovation *and* diffusion of the technology in question. It is nowadays widely accepted that these two phases of technological change are neatly intertwined in many technology areas (see, for example, Rip, Misa and Schot 1995, and Stoneman 1997), and that they should thus be conceptualised as such. Certainly in the case of energy supply this integrated character of innovation and diffusion cannot be ignored.

As a consequence of the inclusion of the diffusion phase, but also due to the fact that energy supply has always been a strongly policy-driven sector, it is necessary to broaden the perspective beyond the analysis of innovation aspects. Therefore, *policy networks* are taken explicitly into account as complementary to *innovation networks*. Both are tied together and supported by a third type of networks: *information networks*. These three types of networks are regarded as underlying and driving the transformation of *large socio-technical systems* (LSTS) such as energy supply.

In order to describe the dynamics underpinning innovation diffusion in a context of structural change, we rely on the concepts of *self-organization, circular causality* and *evolution* (see Chapters 1 and 2). They are central elements for improving our understanding of the shaping of the innovation diffusion pathways.

From a policy perspective, the possibilities and limitations of influencing the innovation diffusion of a specific technology (that is, CHP) during the transformation of a large socio-technical system (that is, energy supply) represent the key research issue. Obviously, this joint analysis of structural change and patterns of innovation diffusion should also deliver some generalised conclusions on the possibilities and limitations of political control.

The argument presented in this chapter draws on case study material on the introduction, improvement and uptake of co-generation technology in the CHP energy supply systems of Germany, the UK and the Netherlands. The innovation diffusion processes differed in many respects and took place in the course of the transformation of the respective regulatory frameworks. The analysis of three different country cases should allow differences and commonalities, both in terms of mechanisms and policy approaches, to be highlighted.

The focus of comparative analysis is on the mutual shaping and co-evolution of the innovation, policy and information networks on the one hand and of their structural or system context within the large socio-technical system of energy supply on the other. In other words, the aim is to identify and compare the importance of different dynamic mechanisms at play between network/micro and structural/macro level of the system under study. These interdependencies between networks and their systemic structural context are crucial in order to be able to take diffusion aspects into account. Not only do properties of the network emerge, but also the structural context of the network is reshaped in the course of the evolution of the network and the innovation diffusion of the underlying technology. While many of the structural features of the system may be independent of the evolution of the network during the innovation phase, they are clearly affected once a wider diffusion sets in.

This chapter is structured as follows. First, the conceptual framework that guided the case studies and the comparative analysis is outlined. Then, a short overview of the main differences between the three countries with respect to CHP provides a rather static picture – but an essential one for understanding the analysis of mutual impacts between networks and system structures, as well as of self-organising mechanisms. In the subsequent section the impact of network evolution and interactions between actors on the transformation of system structures are looked at. Thereafter, the inverse link is discussed; that is, the impact of system structures on interactions in networks and their evolution. This leads on to an analysis of the main reinforcing and delaying (that is, circular) mechanisms that drive system dynamics. In other words, we look into the self-organising features of system change that requires taking a look at the role played by factors in the wider social, economic, political and natural context of the system under study, as well as at historic path dependencies. In each of these sections, material from the three countries is used to illustrate and underpin the argument. Finally, the implications of this analysis for innovation policy are discussed, pointing in particular to the limitations and opportunities of policy to influence the course of change in large socio-technical systems.

CONCEPTUAL FRAMEWORK

The perspective that underpins the research work on the CHP case is based on a number of streams of theoretical and conceptual literature from different scientific disciplines: evolutionary economics, science and technology studies and political sciences approaches to regulation and political control. Of particular interest are those approaches that aim to look at both micro- and macro-level determinants of system change jointly.

The structural, long-term evolution of energy supply as a large (socio-)-technical system has been studied in quite some detail (Hughes 1983; Mayntz and Hughes 1988), with the focus of interest shifting in the early 1990s from the emergence of LSTS to their transformation and reconfiguration (Summerton 1994a). Such studies point to a number of relevant factors that may be at the origin of socio-technical system change: congestion in the physical network of systems, negative externalities, political pressures and ideologies, political developments and contingencies, or changing competitive conditions (Summerton 1994b, p. 12). More recently, the governance of large socio-technical systems has attracted the interest of researchers (for example, Coutard 1999).

In order to address the micro-perspective, the literature on innovation networks and Mode 2 of knowledge production represents a major source of inspiration (Gibbons et al. 1994; Nowotny, Scott and Gibbons 2001), as does the broad range of empirical and theoretical literature from evolutionary economics (see Windrum and Pyka 1999). In order to bring policy more explicitly into the focus of innovation network studies (the 'endogenisation of policy-making'), the role of political determinants and policy networks for innovation has been considered (Weber and Paul 1999). In particular, systems and network approaches from political sciences addressing the issue of political control have been integrated into the conceptual framework for the CHP case (for example, Marin and Mayntz 1991; Görlitz 1994; Mayntz and Scharpf 1995).

Finally, the dynamics of change are interpreted in terms of self-organisation and evolutionary metaphors (see Windrum and Pyka 1999; Chapter 2 in this volume).

In order to make systematic and operational use of these different streams of theory in a comparative study, they have been integrated in a conceptual framework that provides a pattern of analysis for comparison.[1] A central element of this framework consists of the argument that both 'top-down' structural changes (for example, induced by a regulatory reform) and 'bottom-up' innovations (for example, as emerging from networks) can trigger change in LSTS, and thus open up windows of opportunity for new technologies. As these ideal types of driving forces hardly ever occur in an isolated fashion, but tend to exert their influence in parallel, the concept of *co-evolution* of system structures and innovation processes in networks is central to our perspective.[2] Innovations often require contextual adjustments to become successful, and there are several different channels through which such contexts are shaped and transformed. In turn, boundary conditions constrain the innovations that can emerge and diffuse. In the course of the innovation diffusion process, the expectations, concepts and mental framework of the individual actors are also changing; they modify their views on what they regard as technologically feasible and structurally necessary in a continuous learning process. With respect to policy, this interdependent process suggests the hypothesis that neither a top-down nor a bottom-up approach in isolation is most promising but a combined and adaptive approach that takes the issue of co-evolution seriously.

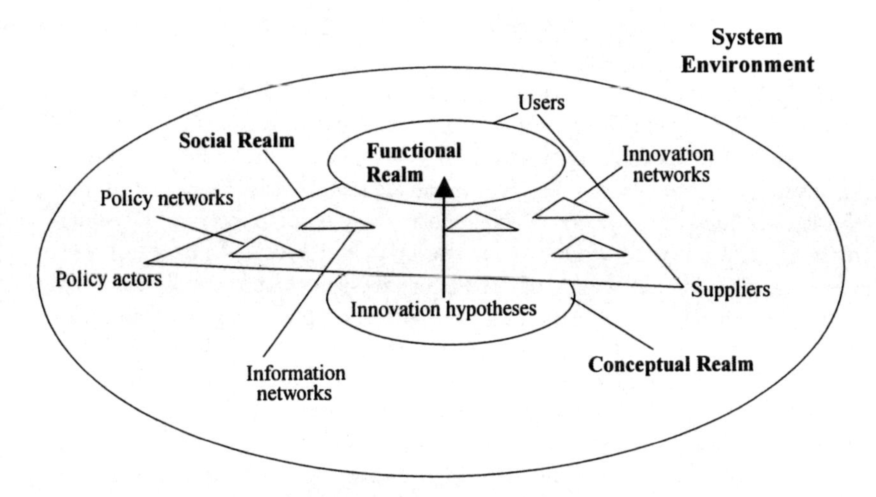

Figure 6.1 Graphical representation of the conceptual framework
(source: Weber et al. 2000)

The argument that networks are driving both the innovation *and* the diffusion of the technology in question has further implications for our conceptual framework. As long as we delimit ourselves strictly to innovation, many elements can be subsumed under the notion of *environment*, that is, as external to the network. Once we consider diffusion processes that affect at least parts of their environment, this environment needs to be split up analytically into a segment that remains fully external, and a segment that is affected and shaped by the innovation in the course of its diffusion.[3] The necessity to define an internal selection environment is the reason to introduce the notion of *system* as an intermediate level between networks and environment (see Figure 6.1).

Here, a broad definition of networks is applied. The interconnectedness and awareness of networks can vary significantly. There can be very close links of which the actors in the network are fully aware, but also loose couplings which some actors would not even regard as network links.

Both innovation and policy networks need to be considered in parallel in order to understand the co-evolution of technological innovation and system structures. Policy networks determine the regulatory and political context conditions for innovation networks, but also the innovation networks themselves lead in the end to a change in structural conditions in industry; for example, as a consequence of the diffusion of new technologies which change the technological structures in which further innovation takes place. Information networks represent a third type of network; they are crucial intermediary organisations that contribute to the shaping of concepts and mental frameworks by providing access to data, assessments and interpretations relevant to the technology in question.

These three types of networks represent the social dimension in this perspective. The second dimension are the realised artefacts, policies, regulations and other structural characteristics that make up the entirety of the energy supply system (including CHP) in the technical sense. Finally, as a third dimension, the concepts, ideas and mental frameworks regarding a technology are explicitly considered ('innovation hypotheses'). Information networks facilitate the exchange of knowledge about new and less conventional concepts, and make sure that they can be fed into policy and innovation networks (see Weber et al. 2000). Similarly, they ensure the exchange of information across these other types of networks.

While the focus is on the interaction between networks and structures, underpinned by the conceptual and mental frameworks of the actors, we should not forget that the entire system considered here is embedded in a wider societal and natural environment. In society we are always dealing with open systems and, as a consequence, there are forces external to the main field of analysis that cannot be ignored. In a systems language, one would say that the external/exogenous forces impinge on structures and networks, but are themselves more or less unaffected by them.

With respect to the dynamics of system transformation, feedback mechanisms or circular causalities are regarded as key concepts. These mechanisms are of a non-linear nature and can thus reïnforce/accelerate, delay/decelerate or stabilise processes of change. In principle, they can operate within and between individual networks, as well as between the different realms of the system under study.

Processes of variation, selection and stability, that is, the typical evolutionary mechanisms, can then be interpreted as being based on combinations of different circular causalities or feedback mechanisms.

While the external environment may constrain system operation and provide, for example, only limited resources, within the framework used here it would not be appropriate to interpret it as a 'selection' environment in the evolutionary sense. In line with the model of self-reference or autopoiesis, selection is regarded as a mechanism which operates from within the system, based on the perceptions of different actors and networks.[4]

With this brief outline of the conceptual framework used for the CHP cases in mind, it is now possible to reformulate the specific research questions addressed in this chapter:

- What are the main factors at micro-/network and macro-/structural level driving innovation diffusion of CHP, and how do they compare in terms of importance?

- What are the main dynamic and reinforcing mechanisms between the two levels, and what is their relative importance in the three countries?
- How effective are micro-, macro- and combined policy strategies in supporting the innovation diffusion of CHP?

A COMPARATIVE VIEW ON KEY CHARACTERISTICS AND DETERMINANTS

A first comparison of the situation and main developments in CHP in the UK, Germany and the Netherlands (see Table 6.1) highlights significant differences in terms of diffusion, but rather minor differences in terms of the technological innovations that were achieved.[5] It shows that the Netherlands has clearly been the most successful example in terms of increasing the contribution of CHP to energy supply. In the UK, quite some progress has been made since liberalisation and privatisation in 1989, but CHP has nevertheless remained a marginal technology. In Germany, a significant capacity has been maintained, and even expanded. Moreover, it is the only one of the three countries where district heating plays an important role.

The technology of CHP does not differ significantly in terms of technical and economic performance from country to country, even if regulatory differences may favour one or other specific type of application. For example, metering equipment for heat and electricity was particularly important in the UK, where the early liberalisation required accurate measurements to underlie economic evaluation. An interesting difference can be observed in relation to service innovations for CHP. In fact, while one of the key innovations that was shared among all countries was the trend towards the packaging of small and medium-sized CHP systems, the extent to which this was combined with a full service option was most pronounced in the UK, somewhat less in the Netherlands, and quite remarkably delayed in Germany. Differences can also be observed in terms of the timing of introducing packaged systems. In the UK, the early liberalisation facilitated the introduction of full service options that boosted the interest in packaged systems. In Germany, the standardisation of design procedures by the influential engineering associations was one of the initiatives that triggered the widespread recognition of small-scale CHP as a reliable and acceptable option.

As the second stage of comparison, we look at the role and importance of the main determinants of innovation diffusion in the three countries. Technically and economically, the differences are rather minor, pointing to the importance of structural and behavioural aspects of innovation diffusion.

At structural level, the largest differences can be observed in terms of the organisational, institutional, regulatory and legal framework conditions. First of all, the degree of decentralisation of energy supply emerged quite clearly as a factor that favours CHP, and in particular also district heating. In Germany, local authorities and utilities are comparatively powerful institutions, which in many cases supported and facilitated the adoption of CHP.

The existence of horizontally integrated energy service companies is another important feature favourable to CHP. While horizontally integrated local utilities have existed in Germany for a long time, they only emerged over the last decade in the UK and the Netherlands. An interesting feature in the UK is that new types of private horizontally integrated energy service companies were created during the 1990s.

Table 6.1 Comparative overview of CHP in the UK, Germany and the Netherlands

	United Kingdom	Germany	Netherlands
Triggering events	1989 (& 1983)	Adaptation policy until 1998 reform	1989 (& 1984) 1998
El. capacity early 1980s	Low (4% of total in 1983)	Medium-high (18% of total in 1980)	Medium (10% of total in 1983)
Types of applications	Mainly industrial, some small-scale	Industrial, district heating, small-scale	Mainly industrial and small-scale
Energy policy strategy	In 1989, radical liberalisation and privatisation of the ESI, little attention to CHP. Later on adjustments of framework to favour CHP a bit more	Until 1998, slow, but continuous adjustment of framework to accommodate CHP, combined with support measures for distant heating 1998: liberalisation	1989: Limited liberalsation combined with targeted incentives for CHP use and an active research policy 1998: full liberalisation
Underlying motivations	Improve economic efficiency of the ESI, later some attention to CO2 problem	Higher efficiency to reduce resource consumption, later on CO2 debates	Efficient exploitation of domestic gas reserve, environment and CO2 debates
Current level of capacity	Low-medium (6% in 1998)	Medium-high (18% in 1997)	High (>25% in 1997, >30% in terms of generation)

The reforms of the regulatory frameworks for energy supply have had major impacts on the incentives and conditions for CHP, both with respect to the innovation and adoption of CHP. However, it would be too easy to argue that liberalisation is by definition positive for CHP; the case of Germany shows quite clearly that the existing CHP capacity could be called seriously into question by a highly competitive framework, if the advantages of the technology are not taken into account in the specific formulation of the new framework.

In the Dutch case, different kinds of targeted financial support schemes (subsidies, loans, etc.) accelerated the adoption of CHP. Similar effects could also be observed for district heating in Germany, whereas in the UK financial support schemes were applied only in very few selected cases, such as CHP schemes with renewable fuels.

Environmental regulations, for example, regarding emissions or efficiency, have influenced not only the uptake of CHP, but also the technological development path taken. In addition to the regulations themselves, their detailed implementation differs from country to country, and even within the same country (especially in Germany). Administrative procedure can be handled relatively liberally, or represent a heavy and incalculable burden.

Issues of (political) culture, problem perception, problem solving and policy approach had quite an important impact on CHP and differ significantly in the three countries. While the different types of approach to CHP are actually not dissimilar, the degree to which consensus could be achieved about them does differ in the three case studies. In the Netherlands, something like a consensus opinion could be established and implemented in policy terms, whereas in the UK a fairly extreme liberalisation policy was implemented, overriding widespread opposition.

The ability to create operational decisions, either based on consensus or enforcement, is very much dependent on the political culture in a country. The Netherlands is a good example of a consensus-oriented political culture, whereas the UK is more conflict-oriented, and the German case is characterised by a wide diversity of arenas of debate and decision making. The ability to achieve consensus about the role and support of CHP differed accordingly.

Liberalisation in the Netherlands and Germany, however, was also influenced very strongly by the European policy framework. Germany maintained until 1998 its monopolistic system. However, regulations were continuously adjusted and targeted financial measures implemented to favour CHP, especially district heating. The most active policy push was applied in the Netherlands, where targeted support and research measures for CHP were combined with a first stage of liberalisation in 1989. This combination strongly favoured CHP in the following years. A fully liberalised energy supply framework was introduced in both countries only in 1998 and is now, in many respects, similar to the British approach implemented from 1990 onwards.

At *micro-level*, the visions, assessments and expectations with respect to CHP, but in particular also the patterns of networking, need to be looked at more closely. The existing *visions* are quite similar in all countries, but the structures of support for these visions differ. Especially in the Netherlands, attempts were made to create the convergence of visions and the establishment of a dominant and shared view on CHP.

The perception and assessment of problems arising from the prevailing ways of supplying heat and power influenced the urgency with which a need was seen to introduce environmentally friendlier technologies. In the UK a particularly high degree of importance was assigned to the economic efficiency of the supply system. Only in recent years, during the CO2 debate, environmental issues have been put on the agenda. In the Netherlands and Germany, energy efficiency, concerns about domestic resources and import dependence, and environmental impacts have been the dominant issues in the CHP debates. The fact that the UK was less concerned about security of supply is not surprising in view of its oil reserves. However, in the Netherlands, with its major gas reserves, the concerns about their long-term availability played a very important role in triggering a change in energy policy.

Concerns about the organisational implications of CHP for the energy supply system were of major importance for the actors in the energy supply sector because they feared a further decentralisation of power supply. The same was true in the UK in the pre-1989 era, but was reinforced by negative prejudices against CHP. In the Netherlands, only certain types of CHP applications (district heating) met with resistance. In any case, the building up of a positive reputation among the public as well as among the engineers turned out to be important.

With respect to the *formation of innovation networks,* the regulatory reform strongly changed the interest structures of the relevant actors and changed their attitudes towards CHP. As a consequence new patterns of collaboration between users and manufacturers became possible, including also intermediaries such as utility and service companies.

In terms of the composition and structure of the innovation networks, the German situation is characterised by a combination of local actions by very innovative small newcomer firms, especially for small-scale CHP, and long-standing linkages between the established manufacturers of energy equipment and utility companies in all other market and technology segments which have only recently embarked upon the new small-scale and service markets. In the Netherlands, two key nodes were created to stimulate the establishment of innovation and adoption networks (NOVEM and the Projektbureau Warmte-Kracht PWK), whereas in the UK collaborations were of a more local nature between individual lead users and innovative technology suppliers.

Only in the Dutch case can one speak of systematic attempts to foster the creation of innovation networks. In both the UK and Germany, the initiative was left mainly to industry and other users.

A number of interesting differences can also be observed with respect to the *formation of information networks*. A major role was played by industrial and engineering associations for the dissemination of CHP information in Germany, whereas in the Netherlands and the UK this function was mainly fulfilled by government-led organisations, at least initially. In Germany, government-led information activities were only of importance in the small-scale area through the involvement of regional energy agencies.

Beyond information dissemination, there were also attempts to achieve standardisation through common design standards in Germany, and a brokering function of key organisations in the Netherlands and the industrial associations in Germany. No equivalent could be identified in the UK.

An interesting difference, especially between the UK and Germany, was that in Germany technical and scientific experts played an important advisory role in the policy debates, reflecting the more rational and less political nature of decision making.

The *formation of policy networks* in Germany is characterised by the very influential role of industrial associations. However, in the course of time, many other actors also got access to the political process. This was further facilitated by the federal political system in Germany, with its multiple points of entry. In the Netherlands, attempts were made to involve a large number of stakeholders and actors right from the beginning, in line with the Dutch consensus-oriented political culture. The British case is characterised by a more polarised political process. At least until the early 1990s, access to the political process regarding energy issues was quite difficult for the promoters of less established options such as CHP. In any event, due to the dominance of the liberalisation issue in British energy policy, CHP hardly ever became an important political issue. The difficulties of bringing new themes into the British policy debates are reflected in the minor role that CHP played in energy policy as compared to the other two countries. In both the Netherlands and Germany, it had been recognised early on that CHP is one of the few available options to reduce the environmental impacts of energy supply.

HOW INTERACTIONS IN NETWORKS SHAPE CONTEXT CONDITIONS

This section addresses the issue of emerging properties; that is, one of the central aspects of dynamism related to and framing the innovation diffusion of CHP. In the system-network perspective underlying this study, two levels of emergence need to be distinguished. First, there are emerging properties of innovation, policy and information networks, resulting from the interactions within them. But at a second level, there are also emergent system properties, such as in particular the functional characteristics of the system under study, including, for example, the diffusion patterns of CHP.

Examples of emerging network properties are their identity, or their degree of closure and reflexivity, but also the existence of procedural links (for example, contracts, collaborations, communication, joint vision, mutual awareness, etc.). The life-cycle features of networks could be mentioned in addition as a time dependent and aggregate characteristic of networks. System properties are reflected in the features of the functional realm, that is, in particular the diffusion patterns of CHP, but also the legal frameworks governing the energy supply system. The characteristics of the conceptual realm, that is, the knowledge about energy supply and CHP, could also be regarded as an emergent system property.

Impacts at System Level

A good example of system properties that result from interactions and decisions in networks are the diffusion patterns of CHP and the resulting change in energy production structures. Whereas in the Netherlands a real boom in CHP took place, to the point that now 40 per cent of power generating capacity is based on decentralised, mostly CHP, plants, the uptake in the UK was more moderate. The target of 5,000 MW installed capacity by the year 2000 has almost been reached, but it represents only about 5 per cent of total electricity capacity in the UK. In Germany, CHP has been more widely used for many years – in contrast to the two other countries also in district heating – but total capacity has grown only moderately over the last 20 years to remain approximately at a level of 18 per cent. Since the reform and liberalisation of energy supply that was finally introduced in 1998, there have even been fears that the CHP in district heating will be jeopardised due to the fierce and short-term economic competition in the sector. In the end, diffusion patterns are based on the individual decisions of companies to adopt CHP for their heat and power needs, but these decisions are preceded by an intense interaction process in the innovation networks.

Similarly, the other main types of decisions in the innovation networks, but also the policy and information networks, give rise to system properties. The emergence of integrated energy service companies is the result of business decisions to introduce service innovations in the energy sector.

Innovation decisions and the first applications of CHP set a learning process in motion which in the end led to changes in the regulatory frameworks. Initially, innovation decisions simply bring users, suppliers and other actors together in order to develop and advance a technology with a promising potential. However, these efforts are not made in a vacuum. While at the beginning of the 1980s the main technological elements to produce reliable small-scale CHP systems were basically available, hardly any firm thought about innovating in this area due to the fact that the framework conditions were not conducive. This changed with the first regulatory reforms to facilitate auto-production and triggered innovation efforts. Once the first pilot and demonstration plants in new application areas (for example, hospitals, hotels, etc.) turned out to be successful, the entire perception of CHP changed. Further learning experiments were undertaken and reinforced the process. It also led to the impression that CHP was in need of adjustments of the framework conditions to be successful. In the Netherlands, this led to a targeted regulatory reform that deliberately favoured CHP, but also to the creation of a new node in the information supply structures with the establishment of PWK. In Germany, it was unclear for quite some time under what types of regulations small CHP plants would fall: power generation or combustion plants. Some regulatory authorities went ahead in applying one or the other regulation. Information access and supply was not so much of a problem in Germany due to the strong and recognised role of industrial and engineering associations in that regard. In many cases, the lobbying of interest groups was important to kick off the debate on new framework conditions. In the end, the structural context for energy supply had changed, and a number of changes were made in response to the innovation efforts by individual firms and networks.

Policy decisions and the related interaction processes in networks, for example, regarding the introduction and modification of new regulations or a support programme, change functional features of the system as a whole. It would go too far to argue that in all cases the CHP debate has determined the way in which regulatory frameworks are shaped, but there is no doubt that the discussion has had an influence. In particular, the debates and later decision on the remuneration of autogenerated power turned out to be crucial.[6] This effect of the CHP debate was strongest in the Netherlands, where an active policy was pursued in favour of CHP. In the UK, the impact was rather minor, because the liberalisation philosophy favoured a passive role for government. However, in the second half of the 1990s, CHP was specifically considered in the support schemes for renewables and a CHP target was set. The German case is to be seen somewhere in between the other two cases. First of all, CHP had already diffused quite successfully in industry and in district heating. There had been active support for CHP schemes after the oil price crises of the 1970s. There has been a continuous updating and adjustment of the framework conditions since the mid-1980s when the new, small-scale CHP systems became available. In other words, the consciousness about CHP led to steady changes at system level. Moreover, we can observe many testbeds for modifications of context conditions for CHP at the level of the *Länder* (that is, the individual federal states). Due to delegation of power in implementing and supervising energy supply frameworks there was scope for experimentation.

Emergent Properties of Networks

Emergent properties can also be observed at the level of the individual networks. The interactions in networks can be characterised in different ways, for example, in terms of their reflexivity, closure or their internal selection mechanisms. In general, the networks in support of CHP are of rather a loose nature, be they for innovation, policy or information purposes. The Dutch case certainly shows the tightest links in all its networks, and also the highest degree of reflexivity and consciousness about the role and membership of a network contributing to the uptake of CHP. The relationships between users and suppliers of CHP systems were relatively close during the piloting and trial phase, but this did not necessarily mean any consciousness about being part of a wider network. One could thus speak of 'tacit networks'.

The German case study also shows that the national network structures were quite dispersed due to the federal political systems. In fact, one should only speak of a single network with respect to federal energy policy because in all other examples of networks, there were regional or application-specific sub-networks in place. While this limited the role for shaping the systemic context, it made it possible to generate variety in niches and regions.

An interesting dimension of CHP networks is their composition in terms of expertise and influence. As mentioned before, core actors can be decisive for the creation of information and support networks but they can also dominate and monopolise the debate and determine the validity of arguments. In the UK, for example, political and ideological arguments strongly guided the energy policy debate, whereas in Germany scientific advice played an important role in giving credibility to arguments regarding policy.

As to the impact of these different network properties on innovation diffusion, a higher reflexivity seems to be helpful, but it should not be confused with centrality. All networks tended to be comparatively loose, implying that there was no necessity to create a high degree of closure and identity to make the networks effective.

Overall, all three case examples show how a new context was created for CHP as a result of the operation of policy networks. Regulatory and policy innovations were of similar importance to the technological advances that had made new forms of CHP possible. New system properties thus emerged which transformed the framework conditions and incentive structures for CHP. The interaction processes that took place in different types of networks for different types of purposes – technological improvements, regulatory adjustments/reforms, creation of information dissemination structures – facilitated the establishment of new and mostly conducive structures and institutions. The three case studies also show that these contexts were quite different, in spite of the fact that the technology is almost identical in all countries. There seem to be limits to the shaping of the context, and these are defined by relationships of power and influence of the main stakeholders, and by the importance that is assigned to the different arguments relevant to the debate (see Appendix 6.1). Finally, the degree of awareness and reflexivity as examples of (emerging) network properties did not really matter in the CHP case. On the contrary, the emergence of fairly strong and influential key network nodes was important for standard-setting and political influence. This centrality needs to be balanced with the degree of dispersion of the network that is needed to generate a variety of solutions adapted to specific local or regional conditions.

HOW CONTEXT SHAPES INTERACTIONS: THE IMPACT OF SYSTEM PROPERTIES ON NETWORKING

In the preceding section, we have looked at the interactions in CHP networks and how they have contributed to the adjustments and even fundamental changes of the structural context in which CHP is applied. It should be clear, however, that CHP has only been one among various other issues in the fundamental debates about the pros and cons of liberalising and transforming energy supply structures. Moreover, network development with regard to CHP was constrained and shaped by the existing structural context. In this section we will thus examine evidence that supports this argument of the impact of system structures on interactions.

First of all, it is obvious that economic and regulatory framework conditions define incentives and barriers for setting up CHP systems (or not). In fact, until 1983 there was no possibility at all for non-utilities to set up CHP and own generation in the UK. Only with the Energy Act in 1983 (which was not very successful) and the Electricity Act in 1989 did the situation change. In the Netherlands, the legal possibilities and financial incentives were decisive for enabling investment in CHP from the second half of the 1980s onwards. In Germany, there was a long-standing tradition of investing in CHP and district heating systems by public and industrial operators. The regulatory and economic framework conditions were continuously adjusted in a process that was to a significant degree left to the self-organisation capacity of the electricity industry and power consumers.

The most important issues of political debate during the 1990s were the economic conditions for connecting independent plants to the power grid.

Defining the terms of exchange, for example, with respect to buy-back and back-up tariffs, was a difficult terrain for negotiation in all three countries, since the enforcement of wheeling (that is, the transfer of power from generator to end-user to another operator's grid system) has turned into the most hotly debated framework issue for CHP in the last few years, especially in the Netherlands and Germany.

Liberalisation improved the possibilities of setting up independent energy service companies that have an intrinsic interest in the combined generation of heat and power. Competition in power supply has also been paralleled by the horizontal opening up of heat, gas, power and water supply from generation to end supply. This development made it more attractive to supply companies to become engaged in CHP, because they could then fully exploit the potential economies of scope inherent to CHP. Before liberalisation, the horizontal integration of different utility services tended to be difficult, with the exception of German local utility companies.

CHP being essentially a decentralised technology, local actors were best placed to make use of it. This is why CHP was comparatively successful in Germany: it was the only one of the three countries under study where local utility companies, often owned by the local authorities, were horizontally integrated suppliers of heat and power that in many cases took environmental concerns seriously. Today, in a liberalised framework, German local utility companies are confronted with the problem that they are often also horizontally integrated with other local services such as water supply, public transport, swimming pools, etc., raising issues of cross-subsidisation and unbundling. Power and heat supply were traditionally overpriced to compensate for losses in public transport and swimming pools, a practice that is nowadays increasingly questioned. It also created problems of competitiveness for the local utilities in the liberalised fields of activity, calling into question the economic viability of CHP plants that were set up under quite different circumstances in the past.

Another important dimension of the transformation process is the change in economic structures. In response to regulatory reform and liberalisation, new types of enterprise emerged in the three countries and changed the rules of the game. Integrated energy service providers (IESCs) replaced functions of the traditional utility companies, but also proposed to take care of all energy activities on the users' sites. This transformation was most pronounced in the UK, and least in Germany. One could argue that in addition to the obvious time lag in Germany, the German local (and sometimes also regional) utility companies already provided a similar kind of integrated service (though usually not on site), whereas in the UK few such activities existed before 1989. Moreover, the competence profiles of (especially industrial) energy service users is in less need of qualified external services in Germany than in the UK.

The political structures in the three countries differ in many respects and had a major impact on the way policy choices were made. Access to the political process, and thus the possibility of bringing new, non-conventional options such as CHP into the debate depends on the openness of the political process. Whether policy networks are formed is also a matter of the opportunities which actors can perceive to exert influence on political decisions regarding regulatory and framework conditions. British energy policy was, until 1989, defined and implemented by a tacit coalition of the ministry in charge of energy affairs and the leading representatives of the monopolistic energy industries. Only the Conservative government's strong will and ideological belief in liberalisation and privatisation were forceful enough to push through the Electricity Act of 1989 and break up the established network. Since then, the influence of government on energy supply has been significantly reduced, implying that policy networks regarding CHP have to focus their attention on the regulator and possibly on technology policy. In both Germany and the Netherlands the political process is much more open, offering numerous possibilities to feed new ideas, findings and concepts into the political process. In Germany, this situation is reinforced by the federal system in which regional authorities play an important role in the implementation and supervision of energy regulation. The policy networks that were enabled by the political structures in these two countries had a clearly perceptible impact on the role of environmental considerations and of alternative technologies such as CHP in the definition of energy policy.

Finally, the role of technological structures for patterns of networking also needs to be highlighted. Large-scale, centralised power supply and grid control implied that the reliability of the supply system had to be based on large units as well, not on a decentralisation of the risk of failure by relying on small-scale plants. While the failure rate of small-scale plants may be higher, the impact of a failure can be easily compensated for by other plants. This is not the case for large-scale units. These two views represent completely opposed supply philosophies. We can observe a lock-in phenomenon for large-scale, centralised supply because once operating in a fully centralised system, it appears as the only logical consequence to continue along this path in order to exploit further economies of scale and guarantee secure supply by means of large-scale plants. Innovation networks in support of alternative supply philosophies are entirely discredited. This could be observed *in extremis* in the UK, but to a more modest extent also in the Netherlands and Germany. Especially in Germany, the existence of well-functioning and cost-effective decentralised plants made the argument less credible.

While the fundamental decisions about liberalising energy supply were not necessarily dependent on the CHP debate (maybe except for the Dutch case), the development of CHP was strongly dependent on the transformation of framework conditions and structures of energy supply. It could be observed how much structural conditions constrain and shape the formation of innovation and information networks during the transition process. Technological, economic, organisational and regulatory conditions affect what actors regard as feasible, or actually experience as feasible. This is one of the reasons why in the end different trajectories and solutions were followed in the three countries. The context matters, even if the technology is basically the same.

This conclusion holds also for the political structures which define who has or does not have access to the political process. By excluding certain assessment dimensions and the actors who represent them, the opportunities for policy networking can be heavily constrained. Policy networks hardly ever appeared as a promising alternative to influence British energy policy (though for different reasons before and after liberalisation), but they were influential mechanisms in the Netherlands and Germany.

THE DYNAMICS OF SYSTEM CHANGE: REINFORCING MECHANISMS AND PATH DEPENDENCIES

In the two preceding sections the impacts of system structures on patterns of networking have been looked at, and vice versa. The differences between the three countries in terms of innovation diffusion patterns of CHP can at least partly be explained on this basis. However, additional aspects need to be taken into account. First of all, system structures and behavioural rules are the result of a historic process which stretches back well beyond the early 1980s which this study took as its starting point. The differences in terms of starting situations at that time need to be analysed as well to understand path dependencies and dynamics. This aspect also relates to the wider context in which the system under study, that is, the energy system, is situated, characterised, for example, by the wider debates on the environment, or the social concerns raised with respect to energy supply.

Second, the main issue of interest here is the interplay between the shaping power of networking on the one hand and the constraining impact of structural conditions on the other. We argue that the actual dynamics of change depend on how these two directions of factors interact, and reinforce or delay each other mutually. In other words, we are looking for circular causalities and examining the extent to which they establish or prevent processes of change.

The Role of History and System Environment

The situation in the three countries was quite different at the beginning of the 1980s. However, in principle, all three countries were confronted with similar problems: the oil price shock was still an important driver of energy policy, and governments tried to reduce import dependency on crude oil. The environmental debate did not yet have a strong impact on the discussions about CHP – a technology which was just about to be rediscovered. What differed significantly was the level of application of CHP: in Germany, CHP was an established technology for district heating and some types of energy-intensive industrial applications. In both the UK and the Netherlands, CHP was of marginal importance only, particularly in district heating. Historically, district heating had not developed in either country due to a lack of carrier organisations at local level. The few district heating schemes that were actually implemented had a bad reputation, though for different reasons in the UK and in the Netherlands. The type of application which is probably most interesting to look at from a comparative perspective is small-scale CHP. It emerged almost in parallel in all three countries, opening up application niches that could not be addressed by CHP before. As regards small-scale CHP the differences in terms of historically shaped structures and conditions were minor between the three countries.

A second key issue concerns the difference between the environment and system under study. In principle, delimiting a system from its environment is an analytical decision taken by the researcher in order to simplify the analysis. The intention is to isolate mutually dependent factors from those that only affect the system under study but are hardly affected by it. Due to the consideration of innovation and diffusion in this case study on CHP, the scope of analysis is obviously quite broad. The environment of a highly complex realm such as energy supply can hardly be interpreted as a 'selection' environment in the narrow sense, but rather as a context that imposes requirements and provides resources. Selection mechanisms operate mainly within the system itself, based on its internal criteria and after interpreting the external constraints accordingly.

With respect to the CHP cases, a number of influential external forces can be mentioned. The oil price shock, while being an energy issue, was an external event which was interpreted in different ways in the three countries. In both the Netherlands and Germany it triggered efforts to use energy more efficiently and raised an interest in alternative technologies. However, especially in Germany, the centralised energy supply was never questioned; most efforts went into more efficient large-scale technologies, whereas in the Netherlands decentralised energy supply became a key issue in the early

1980s. In the UK, the main efforts went into the exploitation of North Sea oil rather than in improving energy efficiency.

A second key external development was the growing awareness about the environmental impacts of energy supply and consumption. It translated into tighter emission limits, with the UK lagging somewhat behind.

A third key external issue was the debate about the EU-wide opening of national borders to energy suppliers from abroad and the liberalisation of energy supply systems in general (CEC 1995). Already anticipated by the UK in 1989, liberalisation became fully effective in Germany and the Netherlands in 1998.

Today, it seems problematic to distinguish national energy supply systems as the main unit of analysis. The process of reconfiguring energy supply systems over the last years has also changed their boundaries. First of all, with the opening of national markets, it is becoming increasingly inappropriate nowadays to look at national systems, even if they still maintain important specificities. Secondly, energy supply firms have started to diversify, for example, in the provision of telecommunications or water supply.

Mechanisms of Change: Interdependence and Circular Causalities Between Networks and Structural Context

Several interesting examples of circular causalities could be identified in the three CHP case studies (see Table 6.2). Their circularity consists of the recursive impact that, for example, a new formal design standard has on engineering practices, which in turn reinforces the influence of this standard. While some of them operate between different networks or between different functional elements of the system, the most interesting examples for the purpose of this chapter are those operating between the level of networking (social realm) on the one hand and the structural context (functional realm) on the other. It is also interesting to consider the two basic directions in which circular causalities can operate. Many of them reinforce the dynamism of change while other circularities tend to stabilise a situation and, in the end, tend to prevent or delay further change by creating rigidities and interdependencies. Circular causalities at structural level in particular tend to be of a stabilising nature with respect to the dominant approach to energy supply. For example, technological and organisational interdependencies in energy supply reinforce each other and prevent change. However, the same type of interdependence can be helpful in speeding up the introduction of new options. Creating technological and organisational interdependencies can thus have a reinforcing effect for the innovation question.

Other circular causalities operate through the perception of functional or conceptual developments by the actors in the social realm. Their reactions to these perceptions can then further affect and shape the issues they have perceived. For example, taking notice of good and successful practices in CHP applications stimulates new attempts to implement CHP that can further contribute to the good practice knowledge base, and so on.

What follows is therefore merely a preliminary set of dynamic mechanisms, which does not claim to be complete. However, the examples listed represent a first attempt to identify empirically candidate mechanisms to be considered in the modelling of innovation diffusion networks.

- *The psychology of examples of success and failure.* Learning from first applications could be observed in many cases, but there were also failed plant projects which undermined any trust in the technology during the early phases. In other words, locally, that is, 50 km around a (technically and economically defined) case of failure of a CHP plant project, the market for CHP was ruined for quite some time.
- *Information diffusion mechanisms* (through institutions such as ETSU in the UK or the industrial associations in Germany) were quite remarkable for CHP; for example, the reinforcing effect which the dissemination of best practice experiences had, also for the creation of practitioner networks. The success of dissemination activities in turn strengthened the support for and influence of the organisations in charge of it, thus facilitating their further dissemination work.
- *Learning by experimenting and piloting* in user-supplier networks, based on the co-operation between suppliers/CHP packagers and lead users, led to the advancement of small-scale CHP technology and enabled the establishment of standardised and reliable designs. In a wider sense, these networks also included regulators and industrial associations that contributed specific knowledge elements. The reasons behind entering into co-operation are manifold: economic arguments can be found as well as environmental interests and long-term considerations regarding the different actors' positions of power in the energy supply system. Successful examples of co-operation attract new interested parties to implement CHP and thus contribute further to learning.
- *Standardisation* effects have been important in all countries, though in different ways. The packaging of plants took place in all countries. It not only reduced costs but perhaps more importantly made the design and implementation of CHP plants sufficiently easy to enable their widespread implementation. In addition, German engineers and their associations played a key role in establishing design standards for small-scale CHP. As long as it was not accepted as a 'standard' technology, it could not really take off. Once its design became standardised into engineering procedures, the acceptance among engineers was almost instantaneous.

One could thus speak of a threshold mechanism due to the standard-setting role of engineering associations.

- Establishment or existence of *carrier organisations*. By creating leading carrier organisations for CHP, Dutch policies established the necessary multiplier platforms. While organisations with a similar function already existed in Germany they did not play an important role in the UK.
- *Learning about new frameworks* for dealing with the new technology in this context is an issue for regulatory authorities as well as utility companies and users; for example, with respect to the import/export of electricity from CHP operators. New incentive mechanisms were at play in the liberalised system that had to be understood with respect to the new opportunities they brought about. The establishment of interconnection standards was an important issue in this respect. The threshold was reached once reliable agreements on tariffs to be paid for exporting power were found.
- *Policy learning* about how to shape the regulatory context was quite a lengthy learning process. It involved at least two major policy networks that represented the dominant energy supply system on the one hand, and the 'new' alternatives on the other. The learning process was based on experiences with CHP in local niches, and on the assessment information (economic, environmental) that became available in the course of the years, both from within the country and from abroad.
- *Consensus building* and the convergence of expectations were particularly important mechanisms in the Dutch political system where they clearly speeded up the introduction of CHP. They operated through the alignment of actors, bargaining deals and the creation of common visions about future energy supply. In Germany, consensus building played an important role within industry itself, though under the threat of political intervention. In the UK this mechanism was not effective.
- *Technological interdependencies* tend to stabilise established forms of energy supply; for example, by sunk costs and economies of scale and scope. Creating new interdependencies of this nature can on the other hand stabilise emerging new options.
- *Economies of scale and scope* also operated in favour of CHP. Economies of scale were enabled by small series production and packaging. They gave rise to cost reductions, but also to simpler interactions with the CHP users who were interested in a simple and easily manageable technology. Economies of scope were particularly important in industries where CHP systems became fully integrated into production processes.
- *Organisational interdependencies* and the mutual reinforcement of economic, technological and political structures are important stabilising mechanisms that hinder change. Until 1989 the UK was an extremely

stable and entrenched system where energy supply companies were closely intertwined with the Department of Energy. On the other hand, once the economic structures and regulatory principles in the UK were broken up, the creation of new interdependencies had a major reinforcing effect on the introduction of new technological and organisational solutions.

- *Learning from good practices* undoubtedly relied on the existence of channels for information dissemination, but the key factor was the feedback from the successful realisation of CHP plants, which led to an improvement in their reputation and provided information about application models from which to learn. This learning from successes thus complements the learning processes within the networks.
- The existence of a *variety of spaces for experimentation* reinforces the process of learning from good practice. In Germany, for example, the decentralisation of political responsibilities offered a great variety of regional spaces for learning due to differences in implementing the federal regulatory framework. The regional practices were then often adopted in other regions.
- Striking the *balance between a still malleable policy and regulatory context* for energy supply and the need to provide reliable framework conditions is another key mechanism at system level. This balance influences the time horizon that actors in the innovation networks can apply, which in turn affects the kind of options they can consider in their R&D and adoption decisions.

Little has been said so far about the timing of these mechanisms. For example, the creation of interdependencies can reinforce the introduction of a new technology, but if these interdependencies are established too early, they may restrict the scope for experimentation and thus for the development of superior solutions. This short review of dynamic mechanisms brings us back to the second of the initial issues/hypotheses of this chapter: identifying the self-organising, evolutionary mechanisms of networking in systems that underlie the innovation and diffusion of new technologies.

The circular mechanisms that could be identified in terms of reinforcing and delaying mechanisms between networks and system context give evidence of self-organising forces at work. At the same time, self-organisation, feedback and circular causalities operate as selection forces differentiating between certain elements that are reinforced and others that are suppressed. Variation, the second key element of evolutionary change, complements this self-organisation process. Self-organisation needs to operate on innovations that have been proposed. In addition, those alternatives that are not selected and further refined in a self-organising way nevertheless contribute to the ad-

Table 6.2 Importance of selected circular causalities in the three CHP country studies

	UK	Germany	Netherlands
Information about successes and failures	+	+	+
Learning through experimenting/piloting	o	++	+
Packaging	+	o	+
Design standards	--	+	-
Existence or establishment of carrier organisations	-	+	++
Learning about new frameworks	+	o	+
Policy learning	+	+	+
Consensus building/shared mental frameworks	-	+	++
Technological interdependencies	++	+	o
Economies of scale and scope	+	+	+
Organisational interdependencies	+	+	o
Learning from good practice	+	+	+
Spaces for experimentation and variety	-	++	+
Reliability vs. malleability of framework conditions	+	+	+

Notes:

-- no impact

- little impact

o moderate impact

+ high impact

++ very high impact

vancement of knowledge and their storage in the conceptual realm. In other words, without variation, there would be no scope for a self-organising process of selecting certain options and refining the shaping of their context.

Policy Approaches and Innovation Diffusion Dynamics

Important differences between the three countries can be identified in terms of the basic policy approaches pursued in the energy field. They are also reflected in the extent to which different types of dynamic mechanisms were set in motion to shape the pathway of CHP. In general terms, we distinguish between 'top-down' structural approaches, aiming to modify the context conditions in the sector in such a way as to stimulate the emergence of more efficient and sustainable solutions, and more 'bottom-up' network-building initiatives, favouring the establishment of new technological solutions through R&D, experimentation and learning in application niches, possibly supported by targeted policy measures. Finally, combined or integrated strategies can be distinguished.

Table 6.3 The role of networking and structural measures for CHP

	Major role of structural driving forces	Minor role of structural driving forces
Major role of network-specific driving forces	*'Co-evolutionary strategy'* CHP in the Netherlands	*'Network-driven strategy'* CHP in Germany
Minor role of network-specific driving forces	*'Strategy of structural change'* CHP in the UK	*'Passive strategy'*

Source: adapted from Weber and Hoogma (1998)

The three types of policy approaches can be roughly assigned to the three countries, as reflected in Table 6.3. The Dutch case turned out to be an example of a combined bottom-up and top-down approach, aiming to improve the framework conditions for CHP while providing targeted support for adoption and innovation. In the UK, policy was based essentially on a top-down, framework-driven approach. It was based on the argument that the liberalised framework would favour the introduction of economically superior solutions. In Germany, prior to 1998 there were many dispersed bottom-up initiatives showing a large degree of variety.

An important question concerns the timing of the policy measures, although it is difficult to discern clear patterns in this respect. For the Dutch situation, one might speak of an early phase until 1989, where mainly bottom-up measures were taken, followed by a first structural reform that was then followed by the 1998 liberalisation. In the UK, it took several years after liberalisation until some relevant adjustments to the framework were made in favour of CHP, but still today there are hardly any effective bottom-up mechanisms at play. In Germany, bottom-up initiatives have always been at play in parallel with minor adjustments of the regulatory framework, at least until 1998. The empirical cases show how critical the importance of a good timing of measures is for the effectiveness of the policy approaches pursued.

ISSUES FOR INNOVATION POLICY: SOME LESSONS TO LEARN?

One of the first questions which is of particular relevance to policy is why many innovation networks actually fail, or why they never make it beyond an early stage. From the CHP case studies, a number of critical points can be highlighted that ought to be taken into account when designing and assessing initiatives to support and facilitate innovation networks. They underline the importance of taking both the role of policy and information networks into account in addition to that of the innovation networks when designing policy strategies for innovation diffusion in large socio-technical systems. This broader view of innovation diffusion processes points to several parallel inroads that can be pursued by policy in order to support the transformation process during different phases of socio-technical change.

Setting up innovation networks alone is often not enough to establish an innovation. There is a high risk that the innovative effort will stop at an experimental stage and will never reach the stage of a wider application. In many cases, this is due to a lack of 'embedding' of the innovations in a compatible structural context. It may require structural and contextual changes to enable the wider uptake of a promising innovation. Regulatory and policy frameworks are critical elements of this context. In fact, policy networking initiatives aiming to establish conducive framework conditions for innovations in a particular field like energy supply can be at least as efficient for creating innovations as the bottom-up operation of innovation networks. So, in the CHP case, structural and regulatory barriers were more constraining than technical and economic ones. The regulatory reforms in energy supply therefore opened up new opportunities for innovative energy supply solutions to emerge, though not necessarily to succeed. Most notably the British case also underlines that a strategy relying solely on transforming the framework does not necessarily lead to the kinds of innovative outputs desired from a policy perspective.

The CHP cases also show how important organisational innovations are to complement the technical dimension of innovation. The existence of potential carrier organisations with an intrinsic interest in the innovation in question turned out to be a decisive factor for establishing innovation networks. Their role is to enable and facilitate the dissemination and learning process with regard to good practice with an emerging technology and thus to establish an information network. This can be set up around an established and respected key actor (for example, an industrial association), but in view of its supporting function in the policy process it may be better to set up an independent body.

Overcoming barriers to both organisational changes and adjustments of the regulatory and policy context requires networks to be in place that promote these political aspects of innovation diffusion in large socio-technical systems. It is important for the success of policy initiatives to take these aspects into account and anticipate them in the design of innovation policy.

Often neglected issues are the shared visions and expectations among the members of a network, because they are essential to help focus the activities of a network. However, especially in the early phases of development they need to be balanced with outsider opinions and a sufficiently broad range of variegated viewpoints within the network in order to test, explore and criticise hypotheses, and thus prepare robust innovations.

Innovation networks can also benefit a lot from leadership or from an 'innovation champion' who helps carry the work over periods of crisis. This leadership function could in specific cases be taken over or at least stimulated by government.

The self-organisation framework as well as the empirical findings point to the effectiveness of a policy strategy that combines structural measures addressing the regulatory and policy frameworks with targeted initiatives at network and technology level. Both these inroads can be supported by the operation of policy, information and innovation networks. Particular attention should be paid to the stimulation of self-reinforcing mechanisms at network and system level and the creation of a variety of novel solutions. Especially the consideration of such 'virtuous circles' should be equally important as the far more common analysis of 'barriers' to innovation and diffusion. Finally, the good timing of measures to support innovation networks is critical. Regulatory and policy reforms can open up windows of opportunity during which innovative technological trajectories can be influenced quite effectively by means of technology policy measures.

Based on the case examples, it can be argued that the Dutch strategy comes closest to the model approach outlined above. Many of the lessons were taken into account and implemented as part of an integrated innovation and diffusion-oriented policy to favour the introduction of co-generation technology.

APPENDIX 6.1: COMBINED HEAT AND POWER – TECHNOLOGY, APPLICATIONS AND EVALUATION[7]

Principle and History

Co-generation or combined heat and power (CHP) can be defined as *the technological principle of simultaneously generating useful heat when generating power*. In conventional power plants only electricity is produced, whereas simultaneously generated heat is – according to the laws of thermodynamics – discharged into the air or into nearby rivers. Heat, either as steam for industrial use or as hot water, is usually generated separately in boiler plants.

The combined production of heat and power is a technological principle which has been used since the beginning of the twentieth century but only in very crude ways as compared to present systems. Industrial applications (for example, in the chemical industry) represented the main share of co-generation, sometimes combined with district heating.

Since the oil price crises of the 1970s there has been a renewed interest in this technology in response to the debates on energy efficiency and liberalisation. The importance of this simple production concept becomes obvious when the thermal efficiency of co-generation plants (~ up to 90%) is compared to separate production of electricity and heat (~ up to 75%).

Main Types of Technologies and Applications

Combined heat and power production systems tend to use components that were originally developed for separate heat and electricity production. In recent years much progress has been made in optimised system integration. The core components can be used as a first classification scheme of CHP systems:

- Steam-turbine based technologies have been applied mainly to large and very large applications (>100 MW).
- Gas-turbine based technologies cover the range of below 1 MW up to about 100 MW. Nowadays, combined cycle steam and gas turbine systems (STEG) are increasingly used.

Reciprocating engines, using gas or diesel, are available in sizes from about 10 kW up to tens of MWs. Small-scale stirling engine and fuel cell systems are still in the pilot phase.

Three main types of functionally different CHP applications can be distinguished, each satisfying heat and power demand by different types of technologies.

- *District or distant heating* for residential heat supply by distributing hot water. District heating systems most often use steam and gas turbine-based technology, now mostly in combined cycles. The operating capacity ranges from 7 MWe up to 250 MWe and more. There are also district heating systems based on waste incineration plants and gas engine schemes for small-scale, local systems.
- *Industrial co-generation*, satisfying heat and steam demands in industry, most often using gas turbines in the range of 2–100 MWe. In case of high temperature steam demand, combined cycle plants in the range of approximately 1 MWe to more than 200 MWe are used. Finally, gas engines are available for industrial co-generation up to 5.5 MWe.
- *Small-scale co-generation*, satisfying residential heat demand of mostly single, stand-alone objects, sometimes also small industrial plants. Small-scale co-generation covers the range of 2–200 kWe, micro systems even down to 0.3 kWe. Gas engines tend to dominate small-scale CHP systems, but currently the application of small gas turbines and fuel cells is under development.

Characteristics and Implications

CHP has a number of characteristics and implications that are relevant for the way it is assessed by different actors. Technological, environmental, business, macro-economic and organisational-institutional implications can matter:

- CHP has significant *advantages in environmental and resource terms* by having the potential of transforming up to 90 per cent of the internal energy of the fuel into useful forms of energy (heat and electricity). In practice, however, and depending on operational conditions of individual plants, the efficiency may be somewhat lower.

- The environmental and fuel efficiency benefits of CHP tend to have a short-term *economic cost* which depends on the size of additional investment costs for CHP (as compared to a simple boiler plant that is combined with power supply through the grid), on operation conditions (long operation hours per year, continuous non-fluctuating loads, etc.), and on connection conditions with the grid (for back-up and export power).
- It is a *decentral technology* because of the constraint to distribute the heat locally. Heat can – in contrast to power – not be easily transferred over long distances, and thus requires local heat consumers. Neither can heat easily be stored.
- Consequently, a wider use of CHP would question the dominant centralised structure of power generation in large-scale power stations.
- Co-generation *links two chains of energy supply* (heat and electricity) and requires a high level of *'horizontal' co-ordination* in order to exploit the benefits of co-production. Usually, the limited flexibility of the heat supply adjusts to operate CHP plants according to the heat load, with power generation being balanced through the grid connection.
- In order to make CHP work, co-operation is required of companies which may not necessarily be willing to co-operate, unless they can exploit and benefit from the economies of scope which are possible with CHP. The crucial issues for *co-operative arrangements* are technical grid connection conditions, and the economic terms of exchange, that is, tariffs for importing and exporting power.[8]
- Large CHP systems (~> 5 MW) are highly *site-specific* and have to fulfil many individual requirements. Smaller ones are becoming more and more standardised but they need to be adapted carefully to the energy supply requirements of the site.
- CHP is not a simple technology; it *requires specialised knowledge* to install and operate it properly, much more than, for example, a simple heat-only boiler (which is – in combination with power supply from the grid – the main alternative to CHP). Dynamic load profiles and the interconnection with the power grid are complex technical, economic and legal issues.
- CHP is not only a more complex way of supplying heat and power to a site, but – with regard to the initial size of investment – also a more expensive one. Hence, it must be seen as an *additional investment*, which is not of core interest for the investor. It must compete with other types of non-essential investment, being hence subjected to relatively strict selection and payback criteria.

- Although CHP provides considerable *environmental benefits* as compared to the traditional electricity-only supply philosophy, it does not necessarily mean a switch to a sustainable path because technically it is just the most efficient continuation of the fossil fuel-based supply path (although it can also be combined with renewable fuels).
- Hence, whereas CHP is technically compatible with the dominant technological paradigm of power supply technology, organisationally it should be seen as representative of a new trajectory of decentralised and co-ordinated generation.

Current and Expected Improvements

Over the last 15 years the most important improvements in reliability and operation costs of CHP systems have been achieved by the pre-packaging of components (and related service innovations), the standardisation of design procedures and system control by remote monitoring. Cooling systems were added to set up tri-generation plants. At component level, higher efficiencies of the prime movers, lower emissions, catalytic converters and the development of small-scale systems at competitive prices had a major impact. Mini- and micro-scale systems were introduced to the CHP market, based on gas turbines and small engines. Fuel cells are expected to revolutionise the CHP market in the coming years, especially at the lower end of the power range.

Alternative Options

Alternatives to CHP can be identified at two levels. First, at macro-level, centralised, large power-only plants with high efficiency (for example, using combined cycles for power generation) can be confronted with decentralised small-scale and horizontally integrated systems based on CHP. Second, at site level, CHP competes with other small- to medium-scale technologies for heat generation (that is, highly efficient boilers, other investments in rational energy use) that are combined with power supply through the electricity grid.

NOTES

1. A detailed description of the framework is given in Weber et al. (2000).
2. On the concept of co-evolution, see Schot (1992) and Weber and Hoogma (1998).
3. Regarding the need to define an internal selection environment, see also the concept of 'Umweltstrukturierung' (structuring of the environment) by Krohn and Küppers (1989).
4. The notion of autopoiesis is used here as reformulated by Druwe (1989) for purposes of analysing processes of political control.

5. The empirical material for the three countries draws on Weber (1999, 2000), for the UK and Germany, and on Arentsen et al. (2000) for the Netherlands.
6. In particular, the tariffs for guaranteeing back-up power in case of emergency and for power exports were hotly debated.
7. This Appendix is based on Weber (1999, chapter 4) where a comparative assessment of different CHP technologies is made.
8. Tariffs have been the subject of numerous disputes in recent years. Electricity tariffs were often non-linear and digressive, thus reducing the incentives for energy efficiency. Moreover, export tariffs tended to be very low. For example, in the mid-1980s, the regional electricity companies in the UK (or boards until 1990) proposed tariffs that only reflected the short-term marginal costs, omitting long-term benefits of lower capacity needs and the 'decentralisation effect.' The latter not only reduces transmission losses due to the vicinity of power source and power load, but also reduces the risk of supply failures and therefore the need for back-up capacity.

REFERENCES

Arentsen, M., P.S. Hofman, E. Marquart and K.M. Weber (2000), 'The case of the Netherlands', in K.M. Weber (ed.), *The Role of Networks for Innovation Diffusion and System Change. CHP in the UK, Germany and the Netherlands*, SEIN Research Report No. 14, pp. 144–99.

CEC (ed.) (1995), *An Energy Policy for the European Union. White Paper*, Brussels: European Commission.

Coutard, O. (ed.) (1999), *The Governance of Large Technical Systems*, London: Routledge.

Druwe, U. (1989), 'Rekonstruktion der "Theorie der Autopoiese' als Gesellschafts- und Steuerungsmodell', in A. Görlitz (ed.), *Politische Steuerung sozialer Systeme*, Pfaffenweiler: Centaurus, pp. 121–43.

Gibbons M., C. Limoges, H. Nowotny, S. Schwartzman, P. Scott and M. Trow (1994), *The New Production of Knowledge: The Dynamics of Science and Research in Contemporary Societies,* London: Sage.

Görlitz, A. (1994). *Umweltpolitische Steuerung,* Baden-Baden: Nomos.

Hughes, T.P. (1983), *Networks of Power. Electrification in Western Society 1880–1930*, Baltimore: Johns Hopkins.

Krohn, W. and G. Küppers (1989), *Die Selbstorganisation der Wissenschaft,* Frankfurt/M.: Suhrkamp.

Marin, B. and R. Mayntz (eds) (1991), *Policy Networks. Empirical Evidence and Theoretical Considerations,* Frankfurt/ M. and Boulder, CO: Campus/Westview.

Mayntz, R. and T.P. Hughes (eds) (1988), *The Development of Large Technical Systems*. Frankfurt/ M.: Campus.

Mayntz, R. and F.W. Scharpf (eds) (1995), *Gesellschaftliche Selbstregelung und politische Steuerung*, Frankfurt/M. and Boulder, CO: Campus/Westview.

Nowotny, H., P. Scott and M. Gibbons (2001), *Re-Thinking Science. Knowledge and the Public in an Age of Uncertainty*, Cambridge, UK: Polity Press.

Rip, A., T. Misa and J. Schot (eds) (1995), *Managing Technology in Society. The Approach of Constructive Technology Assessment*, London: Pinter.

Schot, J. (1992), 'Constructive technology assessment and technology dynamics: The case of clean technologies', *Science, Technology & Human Values*, **17** (1), pp. 36–56.

Stoneman, P. (ed.) (1997), *Economic Policy and Technological Performance*, Cambridge: Cambridge University Press.

Summerton, J. (ed.) (1994a), *Changing Large Technical Systems*, Boulder, CO: Westview.

Summerton, J. (1994b), 'The systems approach to technological change', in J. Summerton (ed.), *Changing Large Technical Systems*, Boulder, CO: Westview, pp. 1–23.

Weber, K.M. (1999), *Innovation Diffusion and Political Control of Energy Technologies. A Comparison of Combined Heat and Power Generation in the UK and Germany*, Heidelberg and Berlin: Springer/Physica.

Weber, K.M. (ed.) (2000), *The Role of Networks for Innovation Diffusion and System Change. CHP in the UK, Germany and the Netherlands*, SEIN Research Report No. 14.

Weber, K.M. and R. Hoogma (1998), 'Beyond national and technological styles of innovation diffusion: A dynamic perspective on cases from the energy and transport sectors', *Technology Analysis & Strategic Managment*, **19** (4), pp. 545–65.

Weber, K.M. and S. Paul (1999), *Political Forces Shaping the Innovation and Diffusion of Technologies: An Overview*, SEIN Working Paper, September 1999.

Weber, K.M., P. Ahrweiler, C. Birchenhall and P. Windrum (2000), 'Conceptual and theoretical framework', in K.M. Weber (ed.), *The Role of Networks for Innovation Diffusion and System Change. CHP in the UK, Germany and the Netherlands*, SEIN Research Report No. 14, pp. 12–26.

Windrum, P. and A. Pyka (1999), *The Self-Organisation of Innovation Networks: An Economist's View*, SEIN Working Paper, October 1999.

PART THREE

Simulation

7. Simulating Innovation Networks

Andreas Pyka, Nigel G. Gilbert and Petra Ahrweiler

INTRODUCTION

Innovation is increasingly recognised as requiring the convergence of many sources of knowledge and skill, usually linked in the form of a network. Today, few innovations can be assigned to a single specific technological field or even a specific firm (for example, Klein 1992). Accordingly, firms cannot expect to keep pace with the development of all relevant technologies without drawing on external knowledge sources. In this respect, today innovation networks are widely considered as an efficient means of industrial organisation of complex R&D processes. In most of the recent research on industrial economics and new innovation theory the increasing complexity of knowledge, the accelerating pace of the creation of knowledge and the shortening of industry life cycles are considered to be responsible for the rising importance of innovation networks (for example, Malerba 1992 and Eliasson 1995). Mechanisms of learning and knowledge creation play a decisive role in the emergence of networks. In this light, networks are to be considered as a component of the emerging knowledge-based society, in which knowledge is crucial for economic growth and competitiveness. In the knowledge-based society not only the quantity of knowledge used is greater but also the mechanisms of knowledge creation and utilisation are changing.

Although recent work in evolutionary economics and elsewhere has examined the role of innovation networks in technical change, it has mainly been at the level of description, introducing for example the concept of national (Nelson 1993) or regional (Cooke and Morgan 1994) innovation systems. It has proved difficult to describe the complex dynamics of innovation networks using conventional methods of analysis (for example, Pyka 1999). Here, we introduce a simulation approach developed by referring to a general theoretical model of innovation networks (Ahrweiler 1999; Gilbert 1999; Gilbert, Pyka and Ropella 2001) and two empirically oriented conceptions of actual innovation networks. We consider innovation networks as evolving from the dynamic and contingent linkage of heterogeneous units each possessing different bundles of knowledge and skill.

INNOVATION NETWORKS

In an influential monograph, Gibbons et al. (1994) argued that knowledge production is in the process of changing from its 'traditional' locus in the ivory towers of academe to being much more closely connected with application contexts. Knowledge production in 'Mode 2' is non-hierarchical and heterogeneously organised. The organisation of knowledge production is flexible, fluent and transitory:

> Examples of this are numerous environmental and agricultural matters, diet and health problems, computerised databanks and privacy. Interactions between science and technology on the one hand and social issues on the other have intensified. The issues are essentially public ones, to be debated in hybrid fora in which there is no entrance ticket in terms of expertise (Gibbons et al. 1994, p. 148).

The economic aspects of innovation have also received increasing attention. Based on the pathfinding work of Nelson and Winter (1982), the Schumpeterian research tradition has been merged with organisational and behavioural elements (especially Simon 1955; Cyert and March 1963) within an evolutionary framework of variation, selection and historical time, in order to capture the dynamics of innovation and their impact on growth, trade and technological change (see Dosi et al. 1988). One of the major motivations behind this work was a discontent with the lack of explanatory power of neoclassical economics in dealing with issues of technological change. In evolutionary economics, the technological element is captured in notions such as 'technological trajectories'; that is, distinct paths of technological development which dominate others and become selected. Several mechanisms have been identified which lead to the establishment of such trajectories. Prominent among these are the mental framework of scientists and engineers, labelled by Dosi (1982) as 'technological paradigms' (see Sahal 1985; Nelson 1987). Other important mechanisms are the persistence of established technological and economic structures or 'lock-ins' into certain technological pathways as a result of a reinforcement of minor comparative advantages or network externalities. In organisational and behavioural terms, evolutionary economics departs from the notion of profit-maximising agents and adopts the concept of 'routines' to describe decision-making processes.

While the earlier work of Nelson and Winter emphasises the market as the main selection environment for technologies, later contributions from a systems perspective pursue a wider approach, and focus on institutional elements as constraining the decision behaviour regarding innovations. Such 'systems of innovation' have been identified, especially at national (see Lundvall 1992; Nelson 1993), but also at regional levels (see Morgan 1997). Within this system context, learning processes among actors are regarded as being crucial, especially those between the users and suppliers and between competitors. From an evolutionary perspective, longer-term, paradigmatic changes in knowledge production are caused by changing patterns of selection in different social spheres: science and technology development, economic development, social changes, institutional factors, and mental frameworks.

The contributions of the sociological literature on innovation and industry dynamics complements the work of evolutionary economics at the micro-level. Network analysis has revealed that new technological innovations are often a 'social construct' rather than, or in addition to, emanating from scientific and technological advances. Network relationships, which complement traditional markets and hierarchies, have become more and more important for the production of knowledge. Systematic efforts using concepts such as 'actor networks' (Callon 1992) or 'socio-technical constituencies' (Molina 1993) have provided initial, if rather static analyses, but do not allow one to study and understand the dynamic behaviour of innovation networks.

Empirical research on the impact of policy measures, especially of regulation, has confirmed the important role played by the political realm in innovation processes. Recent work on policy networks has demonstrated the importance of close interactions between policy makers and technology makers for shaping the institutional and political environment of innovation processes (see, for example, Marin and Mayntz 1991; Mayntz and Scharpf 1995). It recognises the need to look at actor constellations that shape the outcome of policy-making processes, and at the interdependencies between institutional change and actor strategies.

In sum, the economic, sociological and policy literatures have begun to demonstrate that recent developments in knowledge production can usefully be conceptualised in terms of innovation networks. Nevertheless, they still leave several basic questions unanswered. There is no clear definition of what an innovation network is. Rather there are numerous specifications, each emphasising different aspects depending on the perspective of the proposer. Secondly, it is not clear whether there is a single phenomenon applicable to all spheres of innovation, or disparate processes with little or no commonality. Do the innovation networks in biotechnology have the same characteristics as those in telecommunications? Is it useful to treat the processes of knowledge production in the two sectors as similar? Thirdly, the literature is rather silent about the dynamics of innovation networks: how they arise, the growth processes they undergo, and the way they die or merge into other networks.

It is therefore necessary to begin to elaborate a theory of what constitutes an innovation network, together with its dynamics. Here, we make a start by developing a simulation model of a 'generic' innovation network which uses principles of self-organisation and complexity theory (see Chapter 2). The role of simulation in this context is not to create a facsimile of any particular innovation network that could be used for prediction, but to use simulation to assist in the exploration of the consequences of various assumptions and initial conditions; that is, to use simulation as a tool for the refinement of theory. The first part of this chapter is concerned with the development of an abstract simulation model that could constitute a dynamic theory of innovation networks. In the second part, we apply this model to the particular cases of biotechnology and telecommunications to show how the generic theory can be used to illuminate innovation in specific sectors.

The methodology we have adopted accords with Axelrod's (1997) description of the value of simulation:

> Simulation is a third way of doing science. Like deduction, it starts with a set of explicit assumptions. But unlike deduction, it does not prove theorems. Instead a simulation generates data that can be analyzed inductively. Unlike typical induction, however, the simulated data comes from a rigorously specified set of rules rather than direct measurement of the real world. While induction can be used to find patterns in data, and deduction can be used to find consequences of assumptions, simulation modeling can be used to aid intuition (Axelrod 1997, pp. 24–5).

DESCRIPTION OF THE MODEL

The following description of the model is given in a very general form because the simulation platform can be applied to different contexts and can be used to emphasise different perspectives; for example, an economic or sociological perspective on the evolution of innovation networks.

Actors

The starting point for our conceptualisation of an innovation network is the actors. These are mainly firms engaged in R&D. In addition, there are also political actors, venture capitalists, and universities and public research institutes that bridge the gap between applied and basic research. Actors are represented as code that has the standard attributes of intelligent agents (Wooldridge and Jennings 1995):

- autonomy (operating without other agents having direct control of their actions and internal state);
- social ability (able to interact with other agents);
- reactivity (able to perceive their environment and respond to it);
- proactivity (able to take the initiative, engaging in goal-directed behaviour).

The actors in the simulation are able to learn from their own endeavours in research and from other actors with which they collaborate.

Kenes

For the representation of actors we draw on a *genetic* description following the concept of *kenes* (Gilbert 1997). A kene of an actor i ($i \in \{1, ..., k\}$) is a collection of technological capabilities C^j_i in different technological fields ($j \in \{1, ..., l\}$) measured in nominal values where the actor has a certain ability $A^{j,m}_i$ ($m \in \{1,..., p\}$) also measured in nominal values describing its particular specialisation. For each ability the actors develop an expertise level $E^{j,m}_i$ which is a cardinal value measuring the agents' command of the respective ability.

Generally, there are a large number of capabilities, each of which represents a major technological strand; for example, combinatorial chemistry or bio-informatics in the biotechnology case study (Chapter 4 in this volume). There are only a small number of different abilities for each capability, each of which stands for the specific technological trajectory a particular actor is following. Drawing again on the biotechnology case study, two abilities of the combinatorial chemistry capability are serial synthesis and high-throughput screening technologies. The expertise level describes the particular level of know-how and skill an actor has acquired in a certain ability.

For example, a specific capability of an actor i in the biotechnology case study is represented as the triple: (C^2_i; $A^{2,m}_i$; $E^{2,m}_i = 3$), where the second capability stands for combinatorial chemistry, the respective ability m is high-throughput technologies, and the expertise level of the actor is three.

The general structure of a kene is a list of triples (Figure 7.1):

$$\left\{\begin{pmatrix} C_i^j \\ A_i^{jm} \\ E_i^{jm} \end{pmatrix}, \begin{pmatrix} C_i^{j+1} \\ A_i^{j+1,m} \\ E_i^{j+1,m} \end{pmatrix} \cdots \begin{pmatrix} C_1^n \\ A_i^{n,m} \\ E_i^{n,m} \end{pmatrix}\right\}$$

Figure 7.1 Kene

During the innovation processes, this list of triples is modified according to the particular research strategy an actor chooses, as described below. Before discussing the different strategies available for the actors in the model, we shall introduce the way an innovation can be created from an actor's kene.

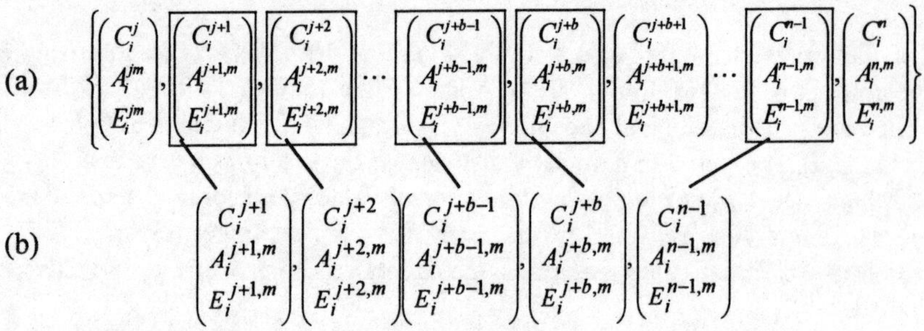

(a) $$\left\{\begin{pmatrix} C_i^j \\ A_i^{jm} \\ E_i^{jm} \end{pmatrix}, \boxed{\begin{pmatrix} C_i^{j+1} \\ A_i^{j+1,m} \\ E_i^{j+1,m} \end{pmatrix}}, \boxed{\begin{pmatrix} C_i^{j+2} \\ A_i^{j+2,m} \\ E_i^{j+2,m} \end{pmatrix}} \cdots \boxed{\begin{pmatrix} C_i^{j+b-1} \\ A_i^{j+b-1,m} \\ E_i^{j+b-1,m} \end{pmatrix}}, \boxed{\begin{pmatrix} C_i^{j+b} \\ A_i^{j+b,m} \\ E_i^{j+b,m} \end{pmatrix}}, \begin{pmatrix} C_i^{j+b+1} \\ A_i^{j+b+1,m} \\ E_i^{j+b+1,m} \end{pmatrix} \cdots \boxed{\begin{pmatrix} C_i^{n-1} \\ A_i^{n-1,m} \\ E_i^{n-1,m} \end{pmatrix}}, \begin{pmatrix} C_i^n \\ A_i^{n,m} \\ E_i^{n,m} \end{pmatrix}\right\}$$

(b) $$\begin{pmatrix} C_i^{j+1} \\ A_i^{j+1,m} \\ E_i^{j+1,m} \end{pmatrix}, \begin{pmatrix} C_i^{j+2} \\ A_i^{j+2,m} \\ E_i^{j+2,m} \end{pmatrix} \begin{pmatrix} C_i^{j+b-1} \\ A_i^{j+b-1,m} \\ E_i^{j+b-1,m} \end{pmatrix}, \begin{pmatrix} C_i^{j+b} \\ A_i^{j+b,m} \\ E_i^{j+b,m} \end{pmatrix}, \begin{pmatrix} C_i^{n-1} \\ A_i^{n-1,m} \\ E_i^{n-1,m} \end{pmatrix}$$

Figure 7.2 From an actor's kene (a) to the innovation hypothesis (b)

From the set of triples describing an actor's kene a subset of at least five triples is randomly chosen to represent the particular research focus of this actor. In other words, actors concentrate their research in certain directions and therefore have to be selective in the particular research direction they choose. This subset of triples is then evaluated to give the co-ordinates of a potential innovation, its innovation hypothesis. This process is illustrated in Figure 7.2, where the five boxed triples were chosen in order to construct an actor's innovation hypothesis.

Whenever such an innovation hypothesis is created, the expertise levels $E^{j,m}{}_i$ of the abilities included in the hypothesis increase by one and the expertise levels of abilities not included decrease by one. This reflects learning-by-doing as well as learning-by-using in those fields the actors are active in, whereas the skills in fields the actors are no longer active in depreciate. Finally, if any triple has an expertise level that has dropped to zero, that triple is deleted from the kene. Figure 7.3 illustrates this processes of learning and forgetting for the exemplary kene used above, where we use numbers to represent expertise levels. The first triple has an expertise level of one and is not in the innovation hypothesis. It is therefore deleted. Triples 6 and 8 are also not in the innovation hypothesis and have their expertise level reduced. All the other triples are included in the actor's innovation hypothesis and their expertise levels are increased.

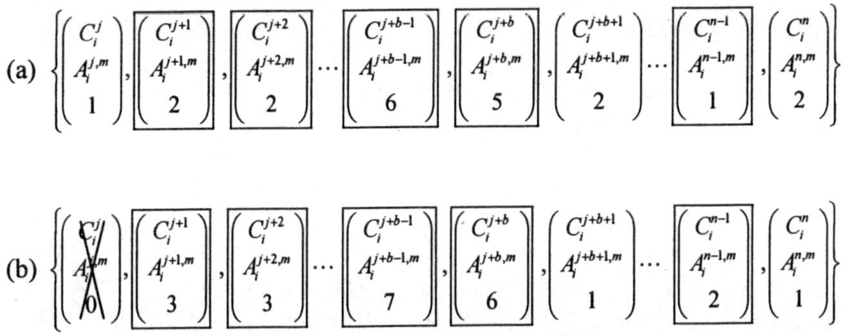

$$(a)\quad \left\{ \begin{pmatrix} C_i^j \\ A_i^{j,m} \\ 1 \end{pmatrix}, \boxed{\begin{pmatrix} C_i^{j+1} \\ A_i^{j+1,m} \\ 2 \end{pmatrix}}, \boxed{\begin{pmatrix} C_i^{j+2} \\ A_i^{j+2,m} \\ 2 \end{pmatrix}} \cdots \boxed{\begin{pmatrix} C_i^{j+b-1} \\ A_i^{j+b-1,m} \\ 6 \end{pmatrix}}, \boxed{\begin{pmatrix} C_i^{j+b} \\ A_i^{j+b,m} \\ 5 \end{pmatrix}}, \begin{pmatrix} C_i^{j+b+1} \\ A_i^{j+b+1,m} \\ 2 \end{pmatrix} \cdots \boxed{\begin{pmatrix} C_i^{n-1} \\ A_i^{n-1,m} \\ 1 \end{pmatrix}}, \begin{pmatrix} C_i^{n} \\ A_i^{n,m} \\ 2 \end{pmatrix} \right\}$$

$$(b)\quad \left\{ \boxed{\begin{pmatrix} C_i^j \\ A_i^{j,m} \\ 0 \end{pmatrix}}\!\!\!\!\!\!\!\!\!\times\ , \boxed{\begin{pmatrix} C_i^{j+1} \\ A_i^{j+1,m} \\ 3 \end{pmatrix}}, \boxed{\begin{pmatrix} C_i^{j+2} \\ A_i^{j+2,m} \\ 3 \end{pmatrix}} \cdots \boxed{\begin{pmatrix} C_i^{j+b-1} \\ A_i^{j+b-1,m} \\ 7 \end{pmatrix}}, \boxed{\begin{pmatrix} C_i^{j+b} \\ A_i^{j+b,m} \\ 6 \end{pmatrix}}, \begin{pmatrix} C_i^{j+b+1} \\ A_i^{j+b+1,m} \\ 1 \end{pmatrix} \cdots \boxed{\begin{pmatrix} C_i^{n-1} \\ A_i^{n-1,m} \\ 2 \end{pmatrix}}, \begin{pmatrix} C_i^{n} \\ A_i^{n,m} \\ 1 \end{pmatrix} \right\}$$

(a) actor's kene at time t, (b) actor's kene at time $t+1$

Figure 7.3 Learning and forgetting represented in expertise levels

The Innovation Oracle and the Entry of New Firms

Innovation hypotheses which, depending on the setting, might represent a new design, a new drug, new knowledge for which a patent application could be made or a new discovery, are submitted each period to an institution we label the Innovation Oracle. It can be understood as a map of a multidimensional and multi-peaked landscape unknown to the actors. The specific coordinates of an innovation hypothesis then determine a point in the landscape. The height of this point indicates the financial reward that can be gained by this particular innovation. When this reward is above a certain threshold, the Oracle allows the innovation hypothesis to succeed. The hypothesis is launched as an real innovation and the reward is paid to the innovator. In order not to provide rewards for the same innovation several times, the landscape is also deformed at and around this particular point. The reward for the point corresponding to the innovation is decreased while the rewards for the immediate neighbourhood are slightly increased, making this region attractive for imitation.

The reward attributed to a successful innovation furthermore signals a promising context for innovation to potential entrants. When the reward is above a certain threshold, a new actor is created as a randomly mutated copy of the successful innovator's kene. By this means, fierce competition is introduced in these regions of the reward landscape. Additionally, an entry process modelled this way supports the diffusion of economically relevant know-how.

Research Strategies

The research endeavours of the actors in the model aim at modifying and improving their kenes in order to discover those parts of the reward landscape which offer at least satisfying rewards; that is, rewards above the threshold below which innovation hypotheses are rejected by the Innovation Oracle. Actors may be endowed with any of three research strategies. The first is the *go-it-alone strategy* where actors engage only in their own R&D and do not collaborate with other actors. The second strategy focuses on imitation only; firms with this *imitative strategy* always search for collaboration partners but do not do any research on their own. And finally we also consider a *collective strategy* where firms engage in their own research as well as in collaboration.

An actor carries out its R&D by incremental research consisting of modifications and improvements within the set of capabilities it has chosen for its innovation hypothesis. Doing incremental research does not, therefore, include any major changes in the sense of substituting capabilities of the innovation hypothesis for capabilities not yet included. Instead, incremental research means exchanging a capability's ability with another ability not yet tried; in terms of kene triples this is done by replacing an existing ability $A^{j,m}{}_{i}$ with a new one $A^{j,m+x}{}_{i}$ chosen from the possible abilities of the triple's capability. The corresponding expertise level of the changed triple is reset to one (Figure 7.4).

The new ability is selected using a simple 'hill-climbing' method of optimisation: if the ability has previously been changed and the effect was beneficial for the success of the hypothesis, then continue in the same direction; otherwise select another ability at random. The actors submit their modified innovation hypotheses to the Innovation Oracle and are thus able to venture into new areas of the reward landscape in order to search for higher payoffs.

(a)

$$\begin{pmatrix} C_i^{j+1} \\ A_i^{j+1,m} \\ E_i^{j+1,m} \end{pmatrix}, \begin{pmatrix} C_i^{j+2} \\ A_i^{j+2,m} \\ E_i^{j+2,m} \end{pmatrix} \begin{pmatrix} C_i^{j+b-1} \\ A_i^{j+b-1,m} \\ E_i^{j+b-1,m} \end{pmatrix}, \begin{pmatrix} C_i^{j+b} \\ A_i^{j+b,m} \\ E_i^{j+b,m} \end{pmatrix}, \begin{pmatrix} C_i^{n-1} \\ A_i^{n-,m1} \\ E_i^{n-1,m} \end{pmatrix}$$

(b)

$$\begin{pmatrix} C_i^{j+1} \\ A_i^{j+1,m} \\ E_i^{j+1,m} \end{pmatrix}, \begin{pmatrix} C_i^{j+2} \\ A_i^{j+2,m} \\ E_i^{j+2,m} \end{pmatrix} \boxed{\begin{pmatrix} C_i^{j+b-1} \\ A_i^{j+b-1,m+x} \\ 1 \end{pmatrix}}, \begin{pmatrix} C_i^{j+b} \\ A_i^{j+b,m} \\ E_i^{j+b,m} \end{pmatrix}, \begin{pmatrix} C_i^{n-1} \\ A_i^{n-1,m} \\ E_i^{n-1,m} \end{pmatrix}$$

Replacing one ability (a) and resetting its expertise level to one (b)

Figure 7.4 Incremental research

The cost of R&D reduces actors' capital stocks. The only way for them to refresh their capital stocks is by successfully introducing a further innovation and receiving the corresponding innovation reward. Actors are also able to choose to do radical research, which means drawing on a new subset of triples for the innovation hypothesis. In deciding between doing incremental or radical research, actors are guided by the success of their record of incremental research as well as by the absolute level of their capital stock.

Exit

The actors leave the simulation when their capital stocks are reduced to zero. Besides the above mentioned costs of performing R&D and the costs of co-operation introduced below, an actor's capital stock is also depreciated by a certain amount each period. Thus every actor has to participate in the innovation process to earn rewards because otherwise their capital stocks steadily decrease to zero.

Collaboration and Networks

Those actors whose strategies also include collaboration can modify their kenes by exchanging knowledge with their co-operation partners. A distinction is made between bilateral collaborations, consisting of a one shot exchange of knowledge, and networks which aim at long-lasting co-operative relationships and a persistent joint modification and improvement of the actors' kenes. Because networks emerge out of bilateral collaboration we begin by introducing collaborations.

At each time step, actors willing to co-operate search for potential partners. In this search process, the qualities of potential partners are displayed through their 'advertisement', a list of the capabilities that they have included in their last innovation hypothesis. In selecting a potential partner the actors remember with whom they previously had a collaboration and generally prefer to renew a collaboration rather than starting a new one with an actor with whom they have no experience. The attractiveness of a potential collaboration partner is measured by the number of capabilities they have in common, because the integration of external knowledge which only has a vague linkage with one's own knowledge is expected to be comparatively more risky and difficult.

If both the actor and the potential partner agree that a collaboration would be beneficial, a new partnership is created. As part of the process of forming a partnership, this leads to a modification of the actors' kenes in the following way. First, all triples from the kene of the collaboration partner are copied into the actor's kene. If the actor already has a capability that is the same as the capability in the triple copied from its partner, the ability from the triple with the higher level of expertise is chosen. Furthermore, the expertise level of the copied triples is set to one in order to reflect the difficulties in integrating external knowledge. Finally, if the triple is in the partner's innovation hypothesis it is added also to the actor's innovation hypothesis. In order to avoid the hypothesis getting extremely long, the merged innovation hypothesis is shortened by randomly deleting capabilities so that it is of a length defined by a normal distribution with a certain mean and standard deviation.

Creating a partnership involves co-ordination costs, which are, however, below the costs of performing one's own R&D. The process of knowledge exchange in a bilateral co-operation is illustrated in Figure 7.5. The innovation hypothesis of actors *i* and *f* are the boxed triples. After creating the partnership with actor *f*, the expertise levels of all triples that appear in actor *i*'s innovation hypothesis are increased by one. Both actors are active in the field corresponding to the first triple's capability, but actor *i* is drawing on ability ${A^{j,m}}_i$ with expertise level one, whereas actor *f* uses ability ${A^{j,m+x}}_f$ at the higher expertise level four. Actor *i*'s ability is therefore replaced by *f*'s and the expertise level is set to one. Also triples from actor *f* now appear in *i*'s kene: $\begin{pmatrix} C_{i(f)}^{j+c-1} \\ A_{i(f)}^{j+c-1,m+x} \\ 1 \end{pmatrix}, \begin{pmatrix} C_{i(f)}^{j+c} \\ A_{i(f)}^{j+c,m} \\ 1 \end{pmatrix}, \begin{pmatrix} C_{i(f)}^{j+c+1} \\ A_{i(f)}^{j+c+1,m} \\ 1 \end{pmatrix}$. The triple $\begin{pmatrix} C_i^{n-1} \\ A_i^{n-1,m} \\ 1 \end{pmatrix}$, which was on expertise level one and which does not appear in the innovation hypothesis, becomes obsolete in the period after the partnership was founded. Finally the innovation hypothesis of actor *i* is augmented by two new triples taken from the original innovation hypothesis of its collaboration partner *f*.

It follows from this procedure that two actors that create a partnership will always submit identical innovation hypotheses to the Innovation Oracle in

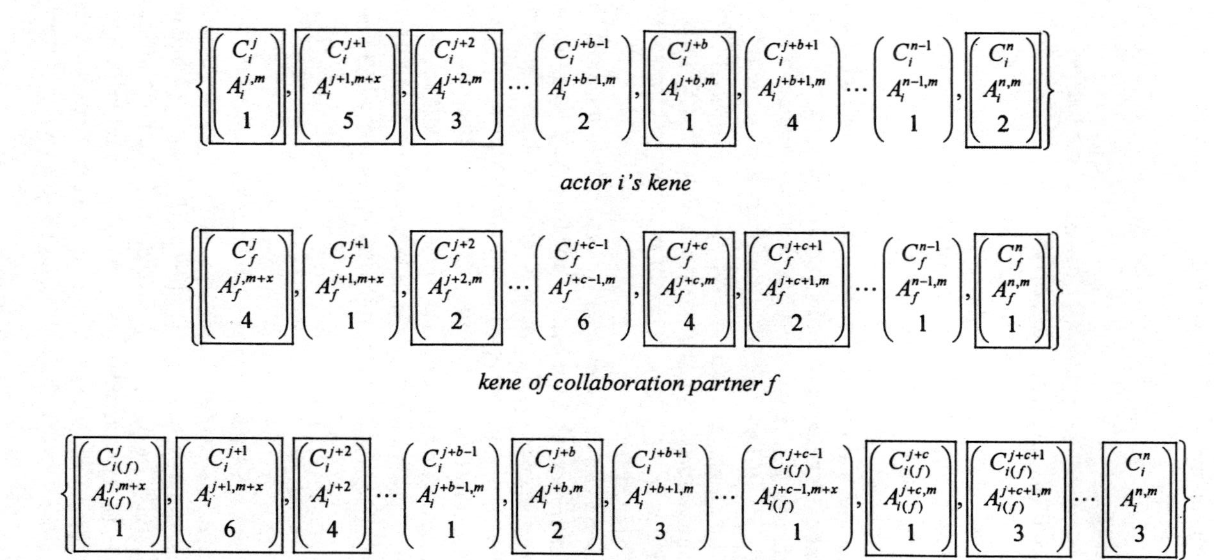

Figure 7.5 Knowledge exchange in a bilateral co-operation

the period after they exchanged their know-how. If the hypothesis results in a successful innovation, the two actors divide the reward between them, in proportion to their capital stock (so that the richer partner gets the majority of the reward). A partnership is maintained for a single time step, although it may be recreated by the partners at the beginning of the next time step.

Networks

Networks can evolve out of partnerships. Networks, in contrast to partnerships, are persistent and can involve more than two actors. A network therefore has an emergent identity that has the effect of protecting the network from short-term dissolution and also reduces internal collaboration costs compared to partnerships. Additionally, networks create certain barriers which prevent them growing too large.

The network evolution process is modelled as follows: An actor *i* that is in a partnership with another actor *f* with whom it already has had a partnership in the past invites this actor to become a member of a network. The partner *f* may accept this invitation if it is not already a member of another network. This kernel of a network grows if one of its members invites a further actor with whom it is in a partnership, with whom it has collaborated before and with whom its network partner has had a previous partnership. Because there is a requirement for this mutual coincidence of preceding partnerships, networks cannot grow explosively.

Members of a network can do incremental and radical research and further collaboration in the same way as other actors. However, they share the results of their incremental research, thus gaining collectively from the different research activities that they each engage in. The consequence of knowledge sharing is that all members submit identical innovation hypotheses to the Oracle and divide their rewards in the case of success. A network will continue so long as it does not have a long run of unsuccessful hypotheses. If it does, the network dissolves and the actors return to operating individually.

Summary

Figure 7.6 summarises the basic structure of the model. The kenes of the agents are transformed in case-study-specific ways to generate potential innovations that are evaluated by an Innovation Oracle to assess whether they are successful innovations ready to be exploited The actors obtain information from the Oracle to support their decisions about how to design their future R&D processes. If actors that are willing to collaborate successfully locate a potential partner who is also willing to co-operate, a partnership is

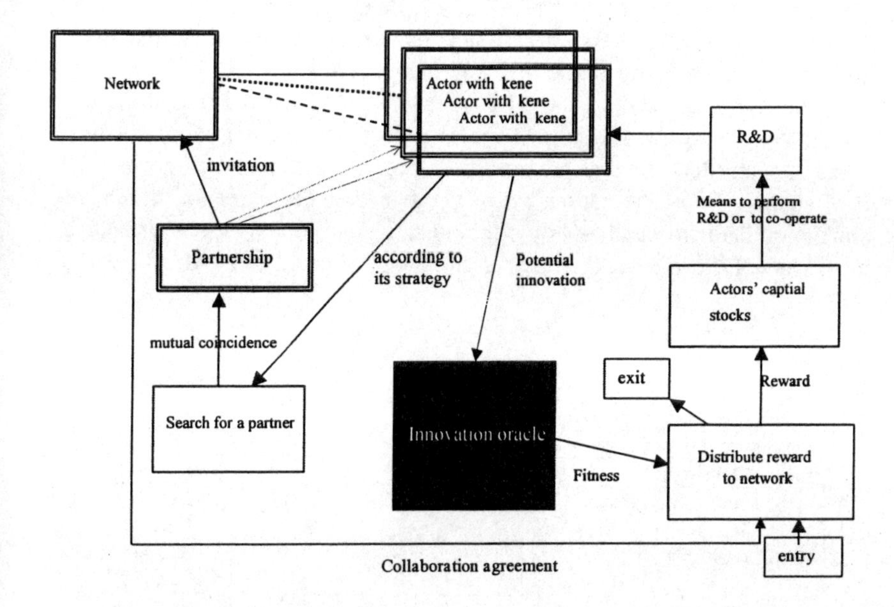

Figure 7.6 The structure of the model

created. From partnerships, persistent innovation networks can evolve comprising two
or more actors. These co-operation processes influence actors' learning, by modifying existing capabilities and creating new ones. When the innovation is successful, the rewards are distributed according to the case study specific rewarding mechanism and the rules of the network concerning the distribution of intellectual property rights, new knowledge, etc. The rewards can then be invested in research. An agent that fails to innovate successfully incurs the costs of research but receives no income and eventually 'dies'.

RESULTS

With the model described above we can perform computational experiments to simulate different innovation networks and their development. We have applied the simulation tool to three case studies, the Virtual Centre of Excellence in personal and mobile telecommunication (Vaux and Gilbert, this volume) (the VCE case), innovation networks in the biotechnology based industries (Pyka and Saviotti, this volume) (the BioTech case) and innovation networks in Knowledge Intensive Business Systems in electronic commerce

(Windrum, this volume) (the KIBS case). In the following paragraphs the results of the VCE and BioTech simulations are described, while the KIBS simulation can be found in Ahrweiler, Gilbert and Pyka (2001).

In the previous section, the basic structure of the model was described qualitatively. However, to implement the design in computer code, it is necessary to introduce several parameters in order to differentiate between the case studies. Experimentation with the range of parameters for which the simulation leads to qualitatively stable results yields a deeper understanding of the model. Moreover, the number and range of parameters which have to

Table 7.1 Parameter settings in the VCE case and the BioTech case

#	Name	VCE	BioTech
1	number of abilities in each capability	10	10
2	maximum level of expertise	10	10
3	form of landscape	peak	peak
4	the number of actors at the start	30	10 LDS 30 DBFs
5	starting capital	10,000	50,000 for LDFs 5,000 for DBFs
6	initial kene length	25	25
7	mean number of abilities required to reproduce an innovation	10	10
8	standard deviation of the number of abilities required to produce an innovation	3	3
9	firm does radical R&D if its capital is below this threshold	2,000	2,000
10	cost of one turn of radical R&D	100	100
11	cost of one turn of incremental R&D	100	100
12	cost per partner per turn of engaging in collaboration	10	10
13	cost per partner per turn of being in a network	8	8
14	how similar two actors have to be in order to link in a network	4	7
15	percentage chance that partner will be selected from friends, rather than the population as a whole	50	50
16	number of actors to try to find one that is compatible for making a partnership	20	20
17	score (reward) that must be obtained from the Innovation Oracle for an innovation to be a success	3,000	5,000
18	number of unsuccessful innovations in a row that causes actors to do radical research	5	5
19	number of unsuccessful innovations in a row after which a network gets dissolved	30	15
20	amount by which the fitness is reduced by a successful innovation	5,000	5,000
21	if the reward received by any actor during a step exceeds this amount, create a start-up	10,000	10,000

be changed in order to get from one empirical case to the other is of interest. In Table 7.1 the parameters of the model are listed for the VCE and the BioTech cases, the grey-shaded areas showing where different parameters are used for modelling the two case studies.

The most important difference between the two cases is the initial distributions of actors. In the BioTech case, there are two very different types of actors: first, Large Diversified Firms (LDFs), for example, the large and established pharmaceutical companies, and second, many small firms which focus on technological competences, the Dedicated Biotechnology Firms (DBFs). Consequently, the two types of firm are modelled with different capital stocks, large for the LDFs and relatively small for the DBFs.

Biotechnology is often characterised as a combinatorial technology where different strands have to be co-ordinated in order to develop a new technology. Additionally, a strong complimentarity between technological and economic capabilities is needed to launch an innovation successfully. Because of the high degree of technical heterogeneity and the high value of the economic assets required, absorptive capacities play a much more important role in the BioTech than the VCE case. In order to learn from a co-operation partner the actors have to share a certain part of their knowledge. The need for a prior shared knowledge base is less important for the actors in the VCE case.

Furthermore, the investment required to make an innovation is smaller in the VCE than the BioTech case. The reason behind this difference can be understood if one considers for example the pipeline of clinical tests a new pharmaceutical device has to pass before it is permitted to be sold. There is nothing equivalent in the VCE case. The economic dependence of the firms on their networks is less in the VCE case and networks dissolve relatively more slowly compared to BioTech, where the actors leave less successful networks more readily in order to look for other opportunities. Finally, the threshold attracting start-ups to enter the market is set higher in the BioTech case, reflecting the high costs of equipment in this field.

The VCE Experiments

After this brief description of the parameter settings relevant for the VCE and BioTech cases, we can now introduce the results of the VCE experiments. We know from the empirical case study that the VCE network includes all relevant actors in the industry and that in the design of the network, policy actors played an important role. Accordingly, the innovation networks that should develop in the simulation experiment should be large and stable and contain similar firms with respect to size and technological orientation.

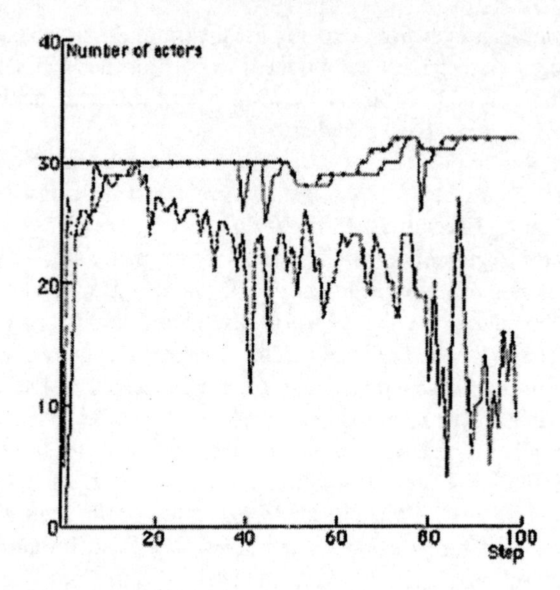

Notes: Dotted line: overall number of firms; solid line: network members; dashed line: firms engaged in partnerships

Figure 7.7 Number of firms in the VCE experiment

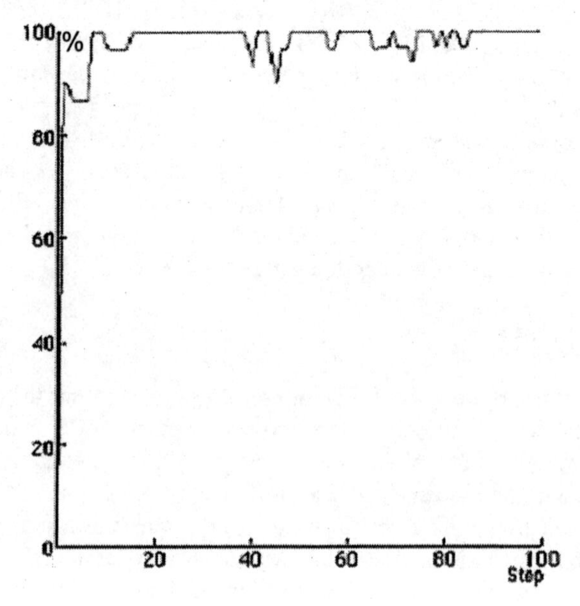

Figure 7.8 Percentage of co-operating firms

Figure 7.7 shows the number of firms (dotted line), the number of firms that are members of a network (solid line), and the number of firms involved in a partnership (dashed line). The number of firms is almost constant during the course of the simulation which is close to what is observed.

The actors engage intensively in partnerships from the start, yielding fast growing networks. This is underlined by Figure 7.8, which shows the percentage of firms that are engaged in some form of co-operation. The percentage increases rapidly at the beginning and remains at almost 100 per cent. It decreases only slightly in periods when the smaller networks are dissolved. Because of the intensive partnering activity, the actors from dissolved networks quickly link to other networks. This also leads to the growth of the average size of a network.

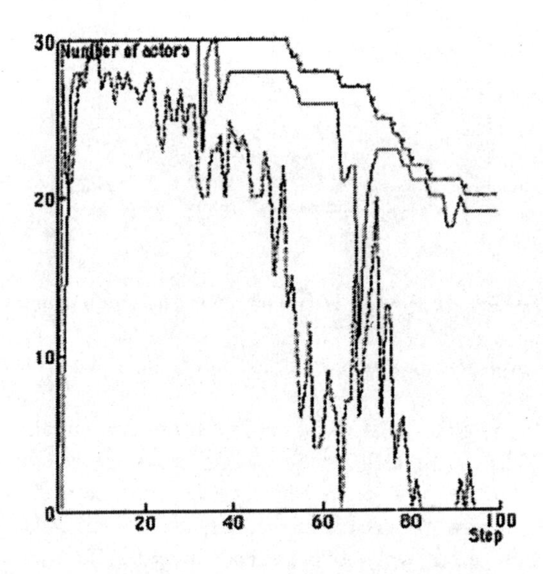

Notes: Dotted line: overall number of firms; solid line: network members; dashed line: firms engaged in partnerships

Figure 7.9 Another VCE simulation experiment

This simulation run does not finish with one big network, but with four: one very large and three smaller ones. The reason for this may be a consequence of an important difference between the simulation and the case study: the omission of an integrating policy actor. The actual development of the VCE network was influenced significantly by a UK Department of Trade and Industry (DTI) policy initiative that has not been modelled. Nevertheless, the tendency of the actors to form large networks is clearly demonstrated in this run.

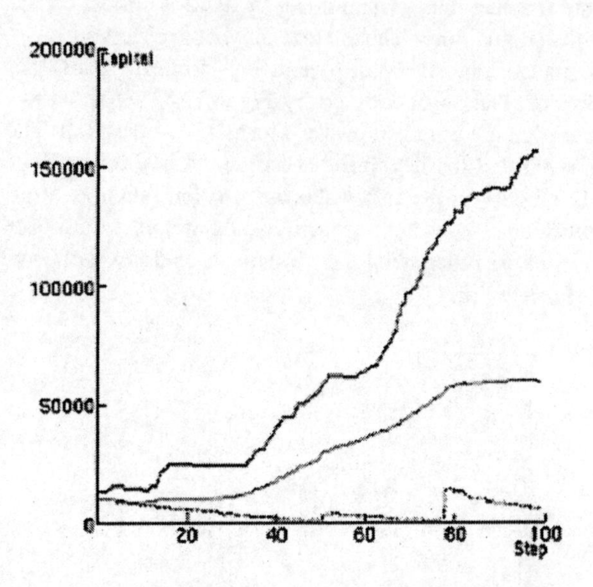

Notes: Dark line: largest firm; light line: average firm; dotted line: smallest firm

Figure 7.10 Development of capital stocks in the VCE experiment

Figure 7.9 shows the results of a simulation run that concludes with only one network. What is the difference with the previous run? In this run, the actors in smaller networks do not engage in cross-network partnerships. When a smaller network dissolves the actors involved have difficulty in joining other more successful networks and, if they fail, 'die'. In this run only those actors able to connect themselves to the large network survive.

Therefore, we conclude that the evolution of a single network as found in the VCE case is possible. But without an integrating actor – a policy initiative – such an outcome is not very likely. However, the networks which emerge spontaneously are large and persistent. This conclusion is also supported by the growth in the size of firms. Figure 7.10 shows the development of capital stocks. As the rewards of successful innovations are shared within the network, the capital stock grows quite steadily. The capital of the largest firm (upper line) is not very different from that of the average firm (medium line). The average is depressed by a small number of new firms entering with small amounts of capital, shown by the plot of the size of the firm having the least capital (lower line). Figure 7.11 shows the capital stock distribution in the final period (step 100). Most firms are about equal in size.

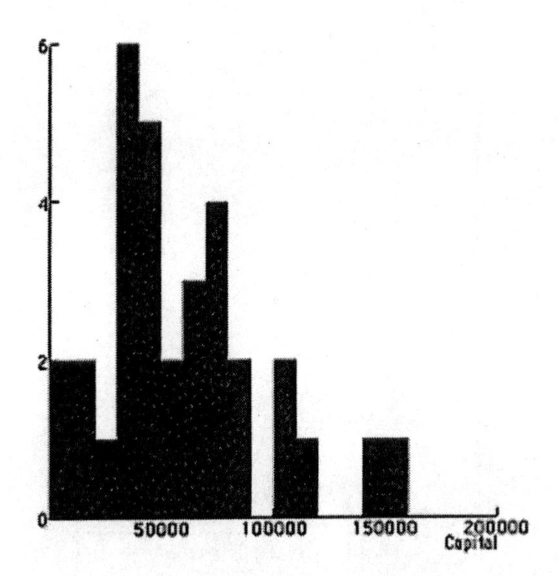

Figure 7.11 Final capital stock distribution in the VCE experiment

Figure 7.12 shows the average (solid line) and maximum (dashed line) rewards for innovations at each time step. In the model, innovation rewards are distributed among the network members according to firm size. This leads to the two firms that received higher rewards in early periods growing more quickly than others – a 'success-breeds-success' effect.

In early periods a slightly increasing trend in the average rewards is evident in Figure 7.12. Knowledge diffuses rapidly among the firms by means of knowledge exchange through networks and partnerships. Therefore, the actors are able to explore the reward landscape quickly. However, the landscape is deformed every time an innovation is successful, reducing the rewards of the firms making innovations in that region. Firms that collaborate with each other increasingly come to resemble one another. The implication is that the overall knowledge base of innovation networks gradually decreases in heterogeneity, making it more difficult to explore new regions. Figure 7.13 shows the gradual decline in the variance of the capabilities over time, a measure of the heterogeneity of the knowledge of the whole population of firms. So, exchanging knowledge over a long period reduces the heterogeneity of knowledge assets, ultimately leading to detrimental consequences for novelty and innovation (Dodgson 1996).

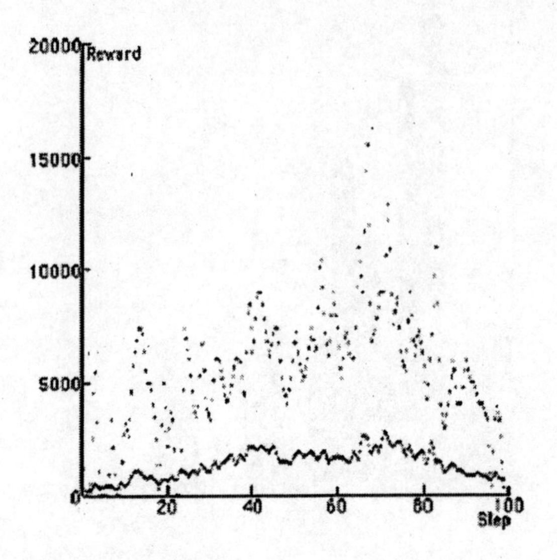

Figure 7.12 Maximum and average innovation rewards

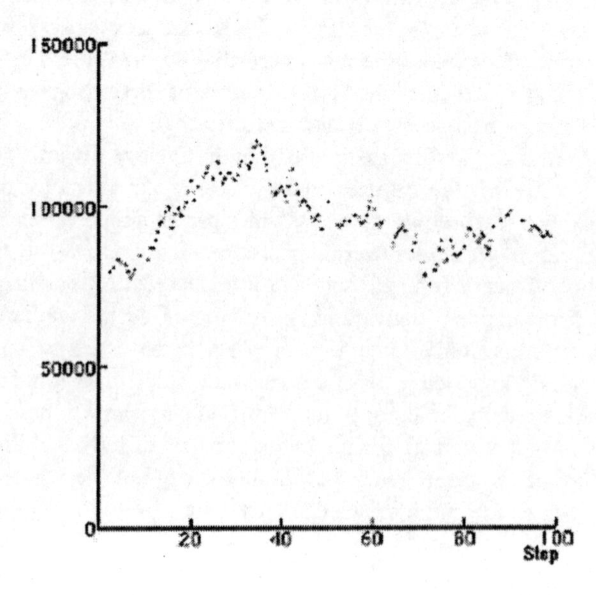

Figure 7.13 Variance of knowledge bases in the VCE experiment

The BioTech Experiments

In BioTech, networks and partnerships are a means of industrial organisation that is used in a flexible way to get access to new knowledge areas. The development of the industry is characterised by a high rate of market entries and exits as well as by a very unequal distribution of firms with respect to size. On the one hand, there are the large pharmaceutical companies and on the other hand there are many small firms specialised in the new biotechnologies. Many small partnerships and networks are observed.

Consequently, the situation to be modelled differs considerably from the VCE case. Figure 7.14 shows the number of firms (dotted line). Only those small firms that are able to find large firms to be partners (solid line) or join a network with a large firm (dashed line) are able to survive this turbulent early period. The first successes in innovation, at around step 50, trigger waves of new firms entering the field.

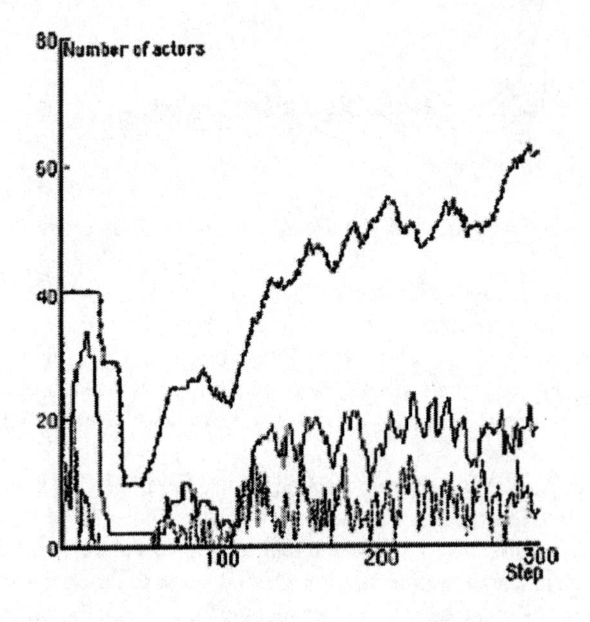

Notes: Dotted line: overall number of firms; solid line: small firms engaged in partnerships; dashed line: small firms in networks

Figure 7.14 The number of firms in the BioTech experiments

Figure 7.15 shows the final distribution of firms according to their capital stocks. The unequal distribution is typical of BioTech industries. There are many small actors – the DBFs – and only a few actors of considerable size – the LDFs.

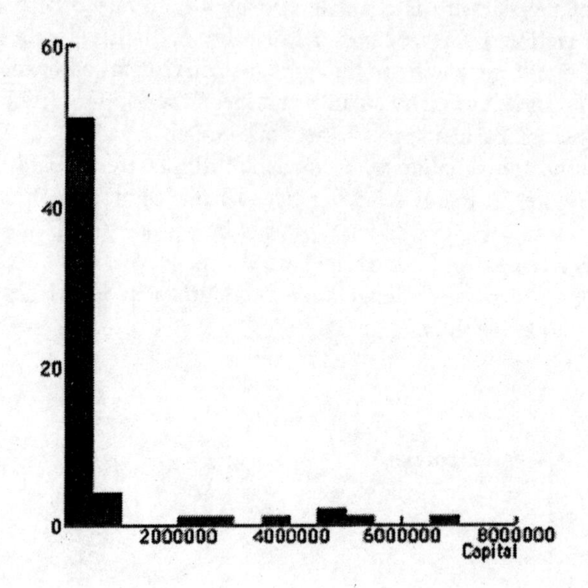

Figure 7.15 Final capital stock distribution in the BioTech case

Figure 7.16 shows the development of capital stocks over time. Whereas the maximum firm size gets larger and larger (light line), the average firm size remains relatively small (dark line). The average remains low because of the continuous entry of small firms. The large firms grow even larger because the innovation rewards earned by networks are distributed proportionately to firm size.

A plot of the percentage of co-operating firms, Figure 7.17, shows a high level of co-operative activity over the whole period. This can be traced back to the unequal distribution of economic and technological capabilities in the population, which is also responsible for the first wave of co-operations (until period 40 or so). However, the development of these collaborations is not accompanied by the successful introduction of innovations. With the first wave of successfully launched innovations (from period 50 to 75) (Figure 7.18), a parallel increase in co-operative activity occurs, which is repeated in the later periods.

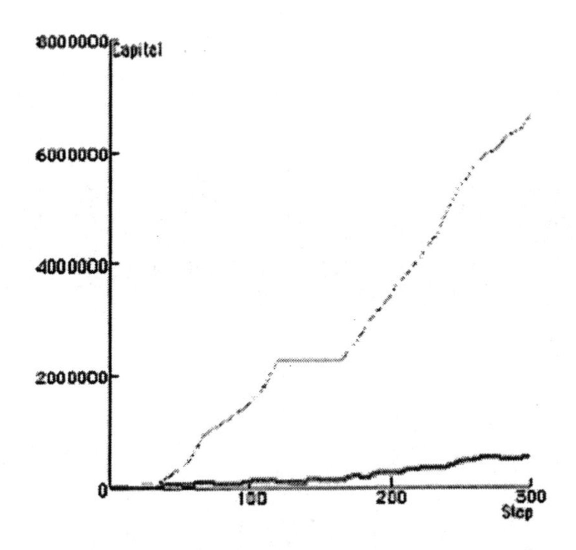

Notes: Light line: largest firm; dark line: average firm

Figure 7.16 The development of capital stocks in the BioTech experiments

Figure 7.18 shows the number of successful innovations as a percentage of the overall number of innovation hypotheses submitted to the Innovation Oracle each period. After the early successes, the actors have to explore new regions in the innovation landscape, which first leads to a decrease in the proportion of innovation hypotheses that are successful. As soon as they discover a new promising region, they start exploiting it jointly, and the number of co-operations therefore increases. In another model (Pyka and Saviotti 2000), we have shown that this can be seen as a change in the role the DBFs play. In early periods they mainly serve as translators introducing the LDFs to the technological opportunities of BioTech. With a wider diffusion of BioTech capabilities, however, the co-operative activities do not stop, but now the DBFs play the role of explorers. This can be seen as a major reason for co-operative activities persisting over time. Accordingly, we find a moderate increase in both networking and partnership activity (see Figure 7.14) caused by positive feedback from innovative success to co-operation.

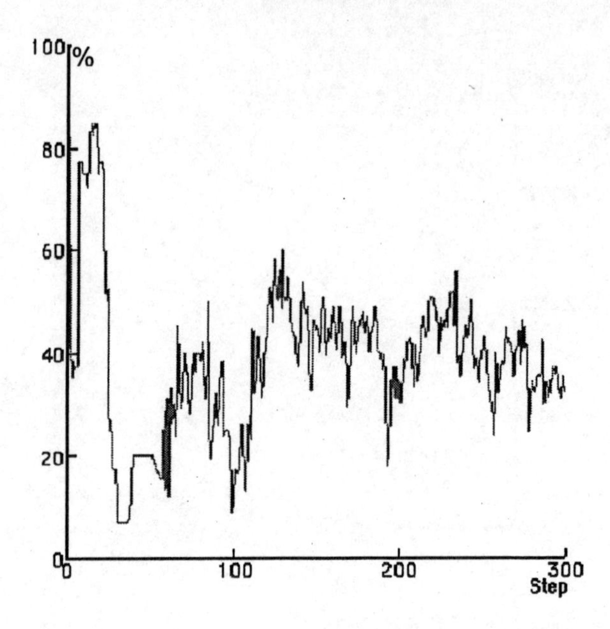

Figure 7.17 Percentage of co-operating firms in the BioTech case

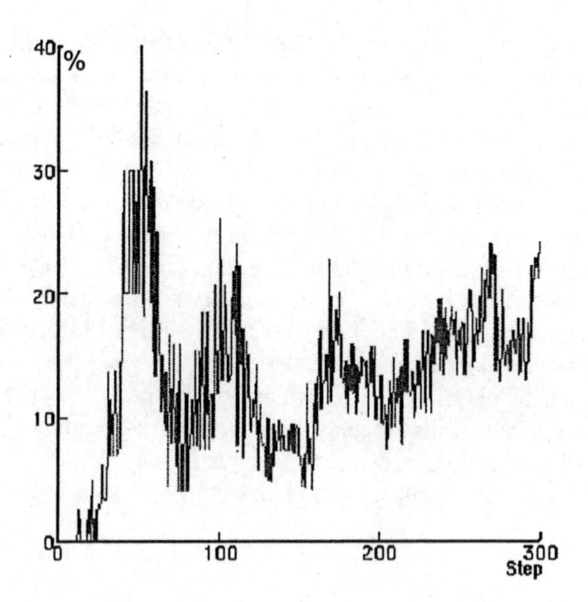

Figure 7.18 Percentage of successful innovations in relation to innovation hypotheses submitted

CONCLUSIONS

The increasing importance of innovation networks in technical change has been emphasised elsewhere, especially in the evolutionary economics literature. However, the processes by which networks are formed, and their role in innovation, is not yet well understood. This is partly because of the complexity of the dynamic processes involved and partly because the actors are heterogeneous and therefore hard to model using traditional techniques. We have shown in this chapter how it is possible to approach these issues through the construction of an agent-based simulation model that allows one to specify, as hypotheses to be tested, the interrelationships between new knowledge, knowledge transfer, selection from the market, and reward structures.

Our prime focus has been on the development of a conceptual basis and a theoretical perspective on the dynamics of innovation network formation. Although we have illustrated the model using two case studies, there remains a great deal yet to be done to test the model adequately. Such testing will involve two rather different approaches:

- The behaviour of the abstract model itself needs to be explored through a sensitivity analysis that will reveal the influence of the model structure. For example, we still need to determine more precisely the circumstances under which the model generates collaborations between actors and therefore yields networks. Under some parameter settings it is possible for all actors to believe that collaborations would be undesirable, and that they would be more effective devoting their resources to their own R&D. Alternatively, under other conditions it is possible for all actors to find the attractions of collaborations to be so strong that the result is a network that includes every actor in the model. Neither of these scenarios is unrealistic. The model provides the opportunity for exploring the circumstances in which these and other scenarios are generated.
- The model needs to be specialised to describe the innovation networks in other sectors (see also Ahrweiler, Gilbert and Pyka 2001).

Finally, it will be possible to draw policy conclusions about the consequences of fiscal and regulatory changes on the propensity for forming networks to encourage innovation in different sectors. The model can be used to examine 'What if' questions to see, in a qualitative way, whether proposed policy changes are likely to have their desired effects.

By investigating such options, by suggesting new ideas for what to try next and testing their outcome immediately, computer simulation can support the policy process using various access points to enter the decision and evaluation areas of policy making (see Ahrweiler, de Jong and Windrum, Chapter 8 in this volume). However, simulation is not only helpful in creating and supporting policy decisions. It can also be used to communicate decisions in a better way. In visualising potential decisions and their possible outcomes by simulation they become more transparent. Results of decision making can be illustrated by showing simulation outputs and, crucially, these illustrations can be reproduced and varied to examine alternatives.

Thus, the decision process as a whole gains a higher degree of legitimacy. Using computer simulation means including modern foresight tools in the decision makers' armoury: it shows that policy institutions do care for consequences of their decision making, that they are prepared to react to changed conditions and outcomes of their target area, and that they search for a way to test and control the quality of their decisions. Using simulation shows that decision makers make good use of all technical support available.

Computer simulation helps to support, communicate and legitimise the policy decision process. In a way, applying IT techniques to the policy process is an auto-logical operation which follows the rationale of technology policy itself, namely to strengthen the linkages between science and users. Here policy makers can apply funded technology to their own area. This feedback or even pay-back between science policy and IT simulation research forms the basic mechanism for a self-organising innovation network.

REFERENCES

Ahrweiler, P. (1999), *Towards a General Description of Innovation Networks. Commonalities and Differences in the SEIN Case Studies*, SEIN-Working Paper, #3, September 1999.

Ahrweiler, P., N. Gilbert and A. Pyka (2001), *Simulation of Innovation Networks. A Conceptual Framework*, SEIN-Working Paper, #16.

Axelrod, R. (1997), 'Advancing the art of simulation in the social sciences', in R. Conte, R. Hegselmann and P. Terna (eds), *Simulating Social Phenomena*, Berlin: Springer-Verlag, pp. 21–40.

Callon, M. (1992), 'The dynamics of techno-economic networks', in R. Coombs, P.P. Saviotti and V. Walsh (eds), *Technological Change and Company Strategy. Economic and Sociological Perspectives*, London: Academic Press, pp. 72–102.

Cooke, P. and K. Morgan (1994), 'The creative milieu: A regional perspective on innovation', in M. Dodgson and M. Rothwell (eds), *The Handbook of Industrial Innovation*, Aldershot, UK, and Brookfield, US: Edward Elgar, pp. 25–32.

Cyert, R.M. and J.G. March (1963), *A Behavioral Theory of the Firm,* Eaglewood Cliffs, NJ: Prentice Hall.

Dodgson, M. (1996), 'Learning, trust and interfirm linkages: Some theoretical associations', in R. Coombs, A. Richards, P. Saviotti and V. Walsh (eds), *Technological Collaboration. The Dynamics of Cooperation in Industrial Innovation*, Cheltenham, UK and Brookfield, US: Edward Elgar, pp. 54–75.

Dosi, G. (1982), 'Technological paradigms and technological trajectories: A suggested interpretation of the determinants and directions of technological change', *Research Policy*, **11**, 147–62.

Dosi, G., C. Freeman, R. Nelson, G. Silverberg and L. Soete (eds) (1988), *Technical Change and Economic Theory*, London: Pinter.

Eliasson, G. (1995), *General Purpose Technologies, Industrial Competence and Economic Growth – With special Emphasis on the Diffusion of Advanced Methods of Integrated Production*, Working Paper, Stockholm: Royal Institute of Technology.

Gibbons M., C. Limoges, H. Nowotny, S. Schwartzman, P. Scott and M. Trow (1994), *The New Production of Knowledge: The Dynamics of Science and Research in Contemporary Societies*, London: Sage Publications.

Gilbert, N. (1997),'A simulation of the structure of academic science', *Sociological Research Online,* **2** (2), http://www.socresonline.org.uk/socresonline/2/2/3.html.

Gilbert, N. (1999), *First Draft of a Model of an Innovation Network*, SEIN-Working Paper, June 1999.

Gilbert, N., A. Pyka and G. Ropella (2001), *The Development of a Generic Innovation Network Simulation Platform*, SEIN-Working Paper, #8.

Klein, B. (1992), 'The role of positive sum games in economic growth', in F. Scherer and M. Perlman (eds), *Entrepreneurship, Technological Innovation and Economic Growth, Studies in the Schumpeterian Tradition*, Ann Arbor, MI: University of Michigan Press, pp. 281–300.

Lundvall, B.A. (ed.) (1992), *National Systems of Innovation. Towards a Theory of Innovation and Interactive Learning,* London: Pinter.

Malerba, F. (1992), 'The organization of the innovative process', in N. Rosenberg et al. (eds), *Technology and the Welfare of Nations*, Stanford, CA: Stanford University Press, pp. 247–78.

Marin, B. and R. Mayntz (eds) (1991), *Policy Networks. Empirical Evidence and Theoretical Considerations,* Frankfurt/M. and Boulder, COL: Campus/Westview.

Mayntz, R. and F.W. Scharpf (eds) (1995), *Gesellschaftliche Selbstregelung und politische Steuerung,* Frankfurt/M. and Boulder, COL: Campus/Westview.

Molina, A. (1993), 'In search of insights into the generation of techno-economic trends: Micro- and macro-constituencies in the microprocessor industry', *Research Policy*, **22**, 479–506.

Morgan, K. (1997), 'The learning region: Institutions, innovation and regional renewal', *Regional Studies*, **31**, 491–508.

Nelson, R.R. (1987), *Understanding Technological Change as an Evolutionary Process*, Amsterdam: North-Holland.

Nelson, R.R. (ed.) (1993), *National Innovation Systems. A Comparative Analysis,* New York: Oxford University Press.

Nelson, R.R. and S.G. Winter (1982), *An Evolutionary Theory of Economic Change*, Cambridge, MA: Harvard University Press.

Pyka, A. (1999), *Innovation Networks in Economics. From the Incentive-Based to the Knowledge-Based Approaches*, SEIN-Working Paper , #1, April 1999.

Pyka, A. and P. Saviotti (2000), *Innovation Networks in the Biotechnology-Based Industries*, SEIN-Working Paper #7.

Sahal, D. (1985), 'Technology guide-posts and innovation avenues', *Research Policy,* **14**, 61–82.

Simon, H. (1955), 'A behavioral model of rational choice', *Quarterly Journal of Economics*, **69**, 99–108.
Wooldridge, M. and N.R. Jennings (1995), 'Intelligent agents: Theory and practice', *Knowledge Engineering Review*, **10**, 115–52.

8. Evaluating Innovation Networks

Petra Ahrweiler, Simone de Jong and Paul Windrum

INTRODUCTION

The increased emergence of innovation networks as the dominant mode of innovative activities requires a radical change in evaluation practices, methods and tools. A new design for RTD evaluation can justly be motivated by this belief. The perspective of innovation networks leads to a new evaluation approach for RTD long-term strategies as implemented within the Framework Programmes of the European Commission which particularly encourage knowledge production in innovation networks. Taking the policy makers of these programmes as the addressees and potential users of a new evaluation approach, this chapter refers to goal attainment questions arising within the European Framework Programmes and is mainly directed to EU policy makers – though it considers national and regional RTD evaluation as well.

In a time of limited public financial resources, politicians and policy makers are increasingly accountable for the expenditure of public money. Although the money the EU spends on science and technology policy is limited compared to national funding rates (it is around 4 per cent of public R&D funding of the EU member states), a shift in RTD policy still needs to be justified to a wider public. Justification of public money expenditure and input into (future) policy making is often fulfilled by evaluations.

The increased importance of the government as financial actor is made particularly evident by the VCE and CHP case studies (see Chapters 3 and 5 in this volume) which confirms the view of some authors who speak of the 'policy framework' of innovation. If the state acts as a supporter and mediator – in short as a network designer co-ordinating science and industry – technology production is extensively funded by public means. This public support causes a high pressure for evaluation: the need to account for expenses, to report failure and success, to raise more money and to legitimise existing policy strategies requires an intensification of evaluation efforts and increases the frequency of their execution.

What makes the state invest in RTD on such a large scale? There are three lines of reasoning for the financial engagement of governmental actors. The most generic reason for legitimising extensive state funding of RTD is that scientific results are public goods which must therefore be paid for by public means. This has, however, been questioned by Callon (1994), who instead proposes rephrasing this argument in a way which anticipates our perspective: in funding RTD, the state invests in the construction, reformation and shaping of techno-economic nets, thus supporting plurality on the level of net formation as well as on the level of options for companies which have to choose technological solutions. A second traditional answer refers to the deficiency of market mechanisms as a reason for the necessity of high state funding: markets normally do not undertake the risk of immaterial investments and do not consider collective and/or long-term interests. The RTD process cannot be divided into segments where investors are able to choose those segments for financing from which most benefits can be expected. The risk and chance components of other parts within the RTD process must likewise be taken into account; often these components do not allow a calculation of the possible success of a funded project at all. Furthermore, private investors cannot completely claim all benefits coming from RTD efforts for their own use in spite of having paid for the actual development. This makes it very unlikely for private investors to get involved in RTD financing on a large scale.

It has been argued that in modern knowledge production the issues of risk, complexity and non-linearity have increased in importance (Kowol and Krohn 1995). In the absence of an alternative financial source, these features require the state to intervene more and more and to fund RTD on an ever larger scale. A third perspective emphasises that there are other important factors for state funding than only compensating for the deficiency of the market. The socio-political tasks of the state require a kind of action which includes RTD as an area of intervention. Criteria for funding do not only take scientific or economic results (output) into account, but also the societal impact of RTD activities. The following societal impact issues are referred to in the European Fifth Framework Programme, for example:

- finishing economic and currency-political integration;
- returning to permanent and lasting economic growth;
- reducing the unemployment rate;
- reducing the phenomenon of social exclusion.

So far, existing accounts of the RTD process have been distinctly partial, mostly focusing on immediate technological or economic outputs. While such partial accounts may have some practicality for a traditional evaluation perspective, they are of increasingly limited benefit to policy makers who want to evaluate the outcome of long-term policy strategies through the results and impacts of particular innovation performances:

> This new policy will also require a review of methods of *ex ante* and *ex post* assessment. Instruments of forward planning and technology assessment will be important as decision-making tools for the allocation of public funds. To make up for deficiencies in innovation selections by the market, governments will need to know beforehand the probable effects on employment, the environment and quality of life of their investment decisions in different areas of R&D. The scope of assessment will broaden and will examine the 'three e's' of efficacy, efficiency and effectiveness in research and innovation policies (Caracostas and Muldur 1997, pp. 21–2).

Evaluation of Efficacy

This relates to questions of goal attainment like: does the project/programme perform as anticipated and fulfil its expected purpose? Within the new modes of technology production we observe a broad variety of choice between various goals and priorities. However, the socio-economic impact of projects and programmes is always at the top of the list. The impacts of all these activities are aimed at reducing unemployment rates, increasing public welfare, enabling democratic developments, enforcing equalising effects, etc. The research outputs alone are not the only target of evaluation; achievement of the broader effects is becoming more and more important.

To evaluate whether a project/programme is attaining its purpose in this latter respect there have to be measures of goal attainment; furthermore these goals must have been previously defined and ranked. Regarding these two requirements (measures, definitions/rankings), the change from output to impact results in a cluster of operationalisation problems for evaluation studies.

> The problems associated with measuring economic impacts pale into insignificance . . . compared to the conceptual hurdles associated with tackling the evaluation of broader social impacts, yet this is precisely the task evaluators and evaluation structures and systems are now being asked to confront (Guy et al. 1998, p. 19).

First, there are basically long-term developments involved; immediate effects can hardly be expected. For evaluation studies this would mean switching from quantitative measuring of outputs to historical and comparative analysis – implying a radical change in tools and empirical approach. For example, it would require not only an *ex post* evaluation but a system of 'operational monitoring' as well.

Second (but related to the first difficulty), it is nearly impossible to draw a direct causal line between the performance of the evaluated unit and its long-term effect on society. Barker and Georghiou (1990) point to the problem of bridging the gap between research funding and its ultimate impact. This is also related to difficulties concerning timing, the attribution of benefits and appropriability.

Third, the impact categories are defined in such general terms that we can only think of very general measures (for example, social indicators) to detect changes – to evaluate the impacts of concrete projects and programmes these general measures need far more refinement and sophistication before they can explain and interpret any positive relations.

Socio-economic impact assessment is going to be the main feature in evaluations of contemporary public policy measures. But socio-economic impact assessment is not self-evident and several writers have recognised this as problematic. Barré (1999) even states that 'our knowledge of the relationships between research activities and society is not up to the task of socio-economic impact assessment through some kind of measurement'. He suggests that measurement should not be regarded as a task of evaluation. Instead, he proposes a new evaluation scheme; that is 'a process of learning and experimentation aiming at building the extended networks which constitute the social systems of innovation' (ibid., p. 9). In this way, evaluation relates scientific activities to the political debate, 'thus adequately serving its users, be they decision-makers, stakeholders or researchers' (ibid.).

Guy et al. (1998) also urge caution regarding the assessment of socio-economic impacts. They argue that there is a need to educate potential customers for evaluation results about the difference between the desirable and the attainable, in order to adjust their expectations to what is achievable with regard to the current state of the art in evaluation theory and practice:

> It is pertinent . . . to point out that the evaluation community is still incapable of satisfying all the information needs of evaluation customers, but . . . this is as much a reflection of unrealistic demands as it is of the state of development of the craft of evaluation (Guy et al. 1998, p. 29).

Evaluation of Efficiency and Effectiveness

Operationalisation problems likewise occur regarding the adequacy of goals, means and instruments (efficiency). Are the long-term impacts mentioned to be achieved by the external design of innovation networks and their technological output? Is RTD policy an adequate means for social politics and its development strategies? Although these are again empirical questions, they are too vague for empirical social science, whether it be sociology, political science or economics. To deconstruct these questions in order to achieve a measure for efficiency is a hard task for transdisciplinary evaluation research. It would require a set of connectivity measures which are not yet available in empirical social research. Once efficiency was proved, one could proceed to questions of effectiveness which refer to the optimal realisation of politics according to targeted results.

As regards evaluation, the basic features of innovation networks impair the ability to reach these evaluation targets. Processes in networks are non-linear and self-organising. While the concrete research goal of a certain network is negotiated during its whole life cycle again and again, the achievement of a planned goal cannot be the focus for the evaluation of its performance. Fulfilling a plan implies linear processes which are non-typical for innovation networks. The additionality of networks points in precisely the opposite direction: the multiple feedbacks and learning results allow one to speak of process benefits of innovation networks, adding to their output benefits. Actors in innovation networks have to codify and translate knowledge to enable access and availability for every network member. The performance of the network is crucially dependent on whether they are successful in doing so. However, who in the evaluation area is interested in these process benefits, which are only indirectly related to 'benefits for the people', as such? Innovation networks stand in need of a customer for their achieved process benefits. Though looking for synergy effects is so far output-related, it does not address the innovation process *per se*.

Identifying these features, our new Network Evaluation Approach discusses the demands of twenty-first century science and technology policy where the European Union aims to contribute to more welfare and well-being in the EU member states. This view is rooted in the idea that the sociopolitical tasks of a (trans)national government require a kind of action which includes RTD as an area of intervention. Collaborative innovation networks are the research settings which are preferred by recent RTD funding strategies. Consequently, criteria for funding not only refer to scientific or economic results (output) but likewise to the societal impact of RTD activities. At the same time, research evaluation has tried to combine the increased call for valid, independent and useful information with the growing public awareness of scientific developments which calls for transparent policy making and, preferably, with the public explicitly represented as one of the stakeholders. Following from this, Georghiou has pointed out that 'if research is put forward as the solution to social problems, . . . it becomes the property of a new set of stakeholders' (Georghiou quoted in Barré 1999, p. 8).

The major impact of extensive policy funding, the related increase of evaluation pressure, the change of the target for evaluation, the resulting operationalisation problems – these are the new features to be captured by evaluation practice with respect to innovation networks and the new mode of technology production. They make a strong claim for a change of perspectives and methods and imply considerable tool innovation within the evaluation area.

THE MAIN CONCERN OF A NEW EVALUATION APPROACH

Whatever reasons can be given for preferring innovation networks as a target for technology funding, the attempt to set them up by policy seems not to take into account the non-linearity of the emerging new structures, which have to be regarded as evolutionary transitions leading to self-organising interaction patterns. The question arises whether the artificial establishment of innovation networks, whether this 'Mode 2 push approach' is successful or even possible. Facing the new modes of knowledge production, science policy makers are confronted with major uncertainties in the areas of planning, foresight and prediction; the establishment, the funding and the evaluation of innovation networks is a difficult target. What options are left for them within the new modes of scientific and technological knowledge production? What are the impact and the consequences of policy strategies targeting innovation networks?

Given that there is no linear control of scientific and technological development by policy, it is nevertheless not sufficient to speak about fragile and uncontrollable processes of growth and decline in self-organising interaction networks. Instead, chances of intervention and reasonable options for policy making must be optimised. These options can be identified by answering the following questions: Can one influence – maybe even predict – the life cycles of interaction networks within 'Mode 2'? Are there preventing and supporting factors with respect to internal or external conditions? Could we imagine construction directives for the setting up of innovation networks? Is there something like an optimal network in terms of the sample of the agents involved, the modes of co-ordination and integration, the timetable, etc.?

Our new Evaluation Approach offers new tools for discovering policy options. This is its main concern. We need new tools to understand the (possible) effects and consequences of policy strategies which try to define, direct, shape and control innovation dynamics. Looking at the current state of traditional approaches to policy evaluation, it becomes evident that the promise to enable orientation, to obtain reliable future scenarios, to anticipate likely forthcoming developments could not be held by the responsible enterprise.

Most studies conventionally use interviews and statistical data to examine the success and failure of policy strategies, but when applied to the area of innovation networks these fail to address the complexity of the target adequately (Pyka 1999; Ziman 2000).

> Networks are formed of heterogeneous units with different capabilities; they develop in environments that are themselves changing as a result of innovations made elsewhere; and the effectiveness of networks depends crucially on bringing together the knowledge and skills of the actors. In short, networks are complex adaptive systems: they are generally self-organising, adaptive to their environment, have no central control mechanisms, and their current state is dependent on their past history. Innovations can be seen as emergent and unpredictable outcomes of the operation of the networks (Gilbert, Ahrweiler and Pyka 2001, p. 2).

What kind of data is relevant to the new modes of knowledge production? How can the various changing parameters and the heterogeneous complex interaction patterns be included? The considerations just mentioned obviously suggest a new methodology for policy evaluation research; and it is above all computer simulation which provides the necessary capacities to contribute here. Informed by case studies which gather probably relevant data, simulation techniques facilitate many types of sensitivity analysis and can be used both in order to answer the questions asked above, and as a tool for scenario construction, guiding the generation of future policy strategies.

Within the Simulation Model (see Chapter 7 in this volume), there are a number of parameters that can be adjusted to study various types of networks. These include the form of the innovation landscape, the costs of networking and carrying out R&D, and the strategies that actors use to choose network partners. The basic set of parameters which can be varied, that is, increased or reduced for special purposes, is shown in Table 8.1.

How can we use the simulation model for evaluation purposes? First, we have to develop a catalogue of policy issues which can be transformed into questions the simulation model might be able to answer. Second, we have to show the connection between these questions and certain parameters or whole set-ups of the model. Third, we have to present some examples for the kind of answers one could expect from such an exercise.

Table 8.1 The parameters of the simulation model

1. the number of abilities in each capability
2. the maximum level of expertise that anyone can have in any ability
3. the maximum reward any innovation can ever obtain
4. form of landscape: one of (:random :peak :bumpy)
5. the number of actors at the start
6. the amount of capital every actor starts with
7. number of elements that make up a kene when it is created
8. the mean number of abilities required to produce an innovation
9. the standard deviation of the number of abilities required to produce an innovation
10. firm does radical R&D in cases where its capital is below this threshold
11. cost of one turn at radical R&D
12. cost of one turn at incremental R&D
13. cost per partner per turn of engaging in collaboration
14. cost per partner per turn of being in a network
15. how similar two actors have to be in order to link in a network
16. percentage chance that partner will be selected from friends, rather than the population as a whole
17. number of actors needed to try to find one that is compatible for making a partnership
18. the score (reward) that must be obtained from the Innovation Oracle for an innovation to be a success
19. the number of unsuccessful hypotheses in a row that causes actors to do radical research
20. number of unsuccessful innovations in a row after which a network is dissolved
21. the amount by which fitness is reduced by a successful innovation
22. if the reward received by any actor during a step exceeds this, create a start-up.

The first task is to operationalise the overall question about the general impacts and consequences of policy strategies targeting innovation in networks which are supposed to be self-organising (see Chapter 2 in this volume). This is a question about the emergence of two interrelated macro-phenomena, namely the formation of networks deriving from the collaboration decisions of single actors, and the successful production of innovations mostly deriving from knowledge exchanges within these networks. Policy cannot force either network formation or innovation production directly (see Chapter 1 in this volume); however, it can target the conditions for the emergence of both macro-phenomena because the micro-level is accessible for policy intervention. This is crucial because on the macro-level there is no central control mechanism and the performance there is completely dependent on the past history (see Holland 1992). The simulation shows how changes on the micro-level (for example, the number of agents involved) affect the emergence of macro-phenomena (for example, the probability of network formation); it also shows how elements (for example, bits of knowledge) and systems (for example, innovation as a system behaviour) connect to one another.

One possibility for categorising policy issues is therefore to collect open questions about the emergence of each macro-phenomenon, network formation and innovation, and later look into their relations. Network formation, to begin with, involves the successful interaction of heterogeneous actors with different capabilities. However, networks, once established, are not stable but are permanently adapting to the changing requirements of their environment, which always shifts to different conditions, reacting to every innovation made by other actors or networks. For policy matters, it would be helpful to know whether there are construction directives for the setting up of innovation networks and whether there is something like an optimal network in terms of the sample of the actors involved, the modes of co-ordination and integration, the timetable, etc. Therefore, policy questions about network formation could address topics like:

- What are the effects of different micro-conditions on the emergence of networks like?
- What are the effects of different actor strategies for partner search?
- What effects do actors' decisions about involvement in networks have?
- What situation on the actors' level generates collaborations and so networks?
- When do all actors join to form one network?
- What are supporting and what are preventing factors for networks?
- How do networks grow and/or die?
- What do the typical life cycles or 'careers' of these networks look like?
- Whether there are certain stages or phases they normally pass through.
- Whether there are points where decline is likely to occur.
- What different types of networks are able to stabilise over time?

The aim of the networks is to produce innovation. The model can show which type of networks, that is, which combination of capabilities within a certain type of environment, is likely to reach this goal. If innovation is the macro-phenomenon under investigation, any questions about what areas of knowledge, what sorts of capabilities, what strategies of learning, etc., are supporting innovation in certain fields contribute to a list of relevant policy issues concerning research agendas. This list includes questions like:

- Should all networks in any special field include complementary capabilities?
- How to encourage incremental innovation without harming radical innovation?
- What is the right blend between innovation and imitation for a certain area?
- What is the impact of decisions about R&D investment in this area?
- What is the effect of different types of learning on the exploration of the innovation space?

These questions now have to be translated into different parameter settings of the model. In addressing, for example, questions about the effects of different micro-conditions on the emergence of networks, we can translate the specifying question about the effects of different actor strategies for partner search into a sensitivity analysis on the parameters 15 and 16. Changing the former will show what impact it has on the behaviour of the model when partners are mainly chosen from the group the choosing actor worked with before (the alternative is that the choosing actor picks partners from the whole population). The latter parameter gives a choice between a 'conservative' and a 'revolutionary' partner search strategy: the first wants the future partner to be very similar to the choosing actor in terms of capability set, the second wants the future partner to be as different as possible. Experiments have indicated that, at the level of the population as a whole, the strategies that individual actors use to select partners are not very important. It does not affect the general story of failure and success: if they use an unsuccessful strategy (and no innovations result from the network), the network and possibly the actors themselves die to be replaced by others, while the more successful networks continue and come to dominate the field.

This example shows how the simulation model would allow the working through of the whole list of policy issues mentioned above. However, this exercise would require a multitude of experiments. The next part of this chapter illustrates the possible use of the simulation model for scenario creation: it reports some experiments with supporting and constraining issues in order to create an 'entrepreneurship scenario' (case study on KIBS, see Chapter 5 in this volume).

THE ENTREPRENEURSHIP SCENARIO

The 'entrepreneurship scenario' describes technological areas like IT where new small and medium companies enter and exit rapidly growing markets as competitors and collaborators in quick succession. The related cycles of innovation networks were observed in the KIBS case study presented in Chapter 5 of this volume (Knowledge-Intensive Business Services in e-Commerce). This case study examined the construction and co-ordination of innovation networks supporting the development and diffusion of e-commerce, focusing on the role played by knowledge-intensive services in connection with their clients. The KIBS study provided a comparison between innovation networks in the Netherlands and the UK, showing the co-evolving dynamics of provider-user relationships in this area.

Important features of the KIBS case as an example for an entrepreneurship scenario must be duly reproduced by the simulation. A summary of the main features of KIBS consists of:

- the appearance of many small-sized networks;
- the high number of market entries (start-ups);
- the unequal capital distribution: a few rich actors (innovators and very successful imitators) and many actors with small capital;
- actors who avoid doing radical research.

In isolating those parameters responsible, according to the KIBS case study material, in order to describe and build a supportive environment for such technological areas, experiments with the simulation model show that the 'entrepreneurship scenario' is sensitive to the following parameters.

Changing the parameter 'number of actors at start', we see that even a small number of initial actors are capable of creating big markets containing many successful actors where numerous small *ad hoc* collaborations take place in a highly competitive environment and many start-ups try to imitate successful firms as shown in Figure 8.1. Starting with 30 actors, we can observe that even after a serious decrease in the actor population (around 30 per cent) invoked by innovation failure, the considerable success of a few actors can produce a 'gold-digging atmosphere' which leads to massive market entries. Accordingly, the entrepreneurship scenario reproduces a discontinuous appearance of new firms in swarms in the sense of Schumpeter (1912).

Changing the parameter which is responsible for the time actors stick to their strategies adds to this effect (here, increase in the time of faithfulness to the strategy). Start-ups in the entrepreneurship scenario are imitators of a successful innovator but are not engaged in radical research themselves – a situation which is simulated through the high threshold value. Start-ups are quite resistant to failure and rather unwilling to switch to other technological options.

While changing the parameter responsible for the degree of similarity between partners at the moment of choice we observe the appearance of many partnerships developing into small networks. A comparably low parameter value means that actors who co-operate may (but need not necessarily) be quite similar. This reproduces the situation in the KIBS case, although in KIBS networks different capabilities and areas of expertise have to be combined to achieve good market results (this mostly concerns differences between technological and economic capabilities). However, far more important is the fact that the difference between KIBS actors does not follow the border between different technological areas. The networks are rather homogenous and simple in their composition. Structurally, they are composed of similar units, namely small and medium sized enterprises (SMEs), all engaged in the e-commerce sector. Building on this general structural similarity, co-operating firms have a mutual value relationship: the activities of and returns to the R&D of an innovating firm crucially depend on the R&D activities of other surrounding firms. This situation leads to many *ad hoc* co-operations in a highly competitive area.

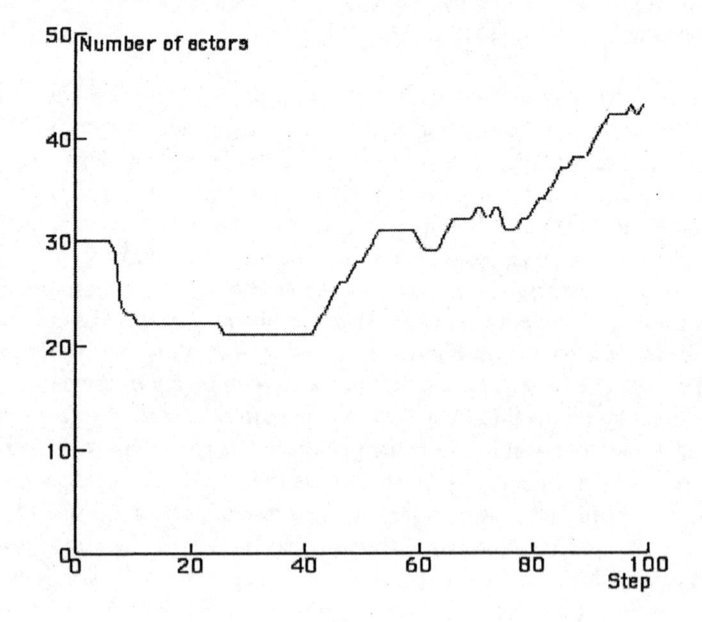

Figure 8.1 Results for number of actors

This competitive constellation is supported by changing the parameter which reduces gains when an innovation is only imitated. The situation that a successful innovation in an area dramatically reduces the profits further to be gained at that point indicates that it is crucial to be first and to be best and, if not, to at least be flexible enough to change innovative strategies and goals quickly. To accelerate the appearance of this effect even further, a new parameter was added to the initial set, namely the 'depreciation costs': they indicate the fixed costs for firms to do their business; that is, motivate them even more to become active in the market as exit becomes more likely.

The example above illustrates how to reproduce by means of computers certain network scenarios guided by the findings of empirical case studies. The big advantage of computer simulation lies in the possibility of isolating single parameters. This enables the user of the simulation to gain access to a complex system which would otherwise be inaccessible for analytical methods. The access gained does not only cover the descriptive power of the model but also its implications for evaluating actions and decisions in the policy area.

According to the results of the simulation, what can R&D policy do to support entrepreneurship scenarios in different technology areas? On the one hand, it is necessary to create acceptable legal and economic conditions for SMEs, especially for start-ups. On the other hand, it is crucial for young firms to be equipped with sufficient initial capital to enter the markets. Policy inputs would have to concern different regulative features:

- financial funding enabling technologically innovative start-ups to enter the respective markets;
- shaping an attractive environment (supportive tax rules, lean bureaucracy, management training, etc.) for young firms to compensate for hazardous conditions in new technological areas.

Furthermore, it was found that the entrepreneurship scenario is highly dependent on a few successful innovators who will take the lead with respect to the 'big wave' of imitators following the first success. For policy makers it would be a big advantage to be able to identify possible innovators who could start the process of building a new market. However, a direct triggering of innovation by finding and supporting the relevant innovators is never possible. Instead, the point in time where an innovation actually takes place becomes crucial. Therefore, our experiments underline not only quite general and well-known possibilities for politicians to spur innovation, but also put emphasis on an important point, neglected more or less so far in the literature. In particular, it is the window of opportunity which is opened up for a short period and which we additionally want to highlight. Then, the policy inputs mentioned above have overall importance. Outside this window in time all policy instruments designed to initiate and further these innovative dynamics are very likely not effective. Accordingly, it becomes crucially important for policy makers to:

- observe the R&D activities and to monitor the innovative space of a technological area;
- develop their own expertise to identify successful innovators early in the process.

THE USE OF (SOFTWARE) INSTRUMENTS IN THE POLICY PROCESS

In Figure 8.2, we point out some access points where computer simulations like the simulation model can enter the policy process as helpful tools. This graphical representation of the policy process leading to the funding of innovation networks, based on the Evaluation and Learning System Approach (ELSA) developed by Technopolis (Guy et al. 1998), describes the four levels of policy context, programme operation, project implementation and participants. (It has been slightly changed here for our purposes.)

Three access points are shown where computer simulations with the simulation model could be used in the way we presented above. The first access point, AP1, concerns the agenda setting process. Simulating ‘entrepreneurship scenarios’ or other relevant market scenarios for different technologies and markets helps to define hierarchies and priorities with regard to the funding of particular research areas. As illustrated in the ‘entrepreneurship scenario’, simulation can show what will happen, for example, if in a new technological area many new market entries take place; additionally, it can immediately show what must be done to get these many entries. Moreover, the simulation model shows how co-operation and networking influence innovation paths and market structures. Addressing questions like this, AP1 supports the agenda setting process of policy making.

AP2 introduces computer simulations like the simulation model at a point where proposals for innovation networks must be evaluated to arrive at funding decisions. With the simulation model it is possible to find out which set of actors is likely to produce sufficient innovation results in a particular area. We can test the composition of innovation networks. The simulation model, for example, shows what different capability combinations and different

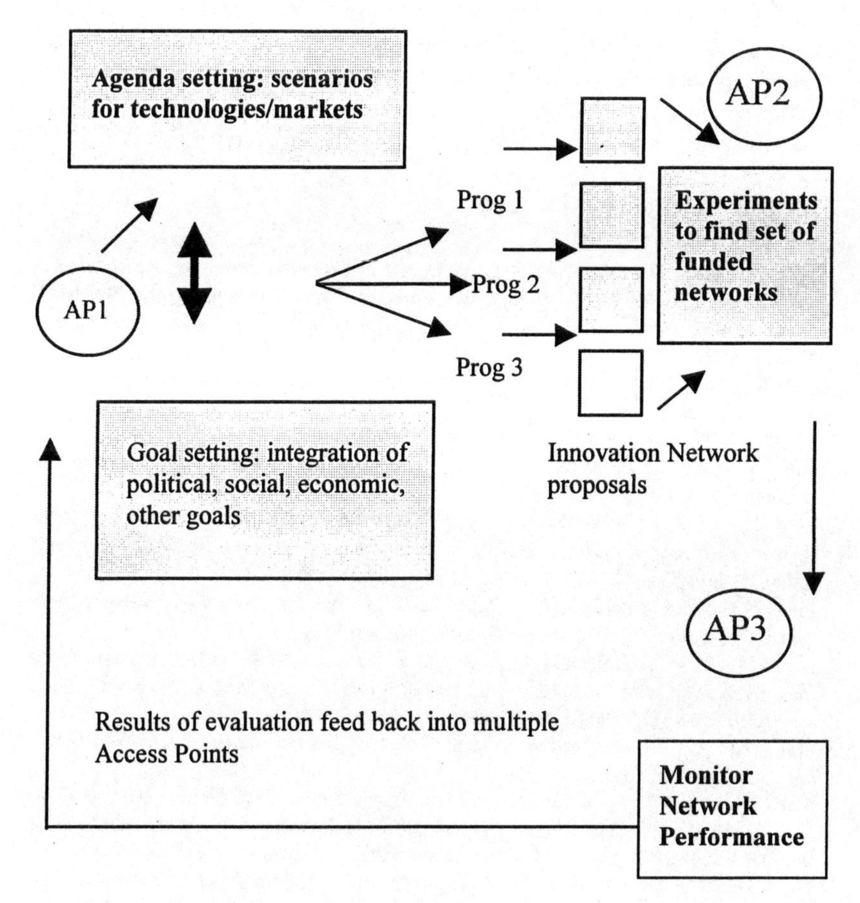

Figure 8.2 Access points for computer simulation in the policy process

numbers of capabilities in networks are likely to achieve and what kind of innovation results.

AP3 is less a point than an area of access: here, the postulated permanent monitoring as an evaluation exercise of policy makers can be supported by computer simulations like the simulation model. The empirical outcome of the funded networks can be compared to the outputs of the simulation. In this process, it is possible to modify expectations, to trace causes as to why expectations were/were not fulfilled and so on. Results from this exercise can be fed back into every part of the policy process.

The use of (software) instruments for policy decision making has been demonstrated by introducing a generic innovation network simulation model which is able to offer a platform for all sorts of policy-relevant experimentation. Work on these tools must be continued, including a 'technology transfer' to the policy area where such instruments could be available to give daily assistance.

REFERENCES

Barker, K. and L. Georghiou (1990), *Evaluation of the Economic and Social Impacts of Publicly Funded Research and Development*, Paper presented to the FORMEZ workshop *Economic Evaluation of Public Projects: A Sectional Approach*, Naples: 8 June 1990.

Barré, R. (1999), *Public Research Programmes: Socio-Economic Impact Assessment and User Needs*, The IPTS Report – Special Issue: *Evaluation and Research Activities*, IPTS – JRC – Seville: December 1999, pp. 5–9.

Callon, M. (1994), 'Is science a public good?', *Science, Technology and Human Values*, **19**, 395–424.

Caracostas, P. and U. Muldur (1997), *Society, the Endless Frontier. A European Vision of Research and Innovation Policies for the 21st Century*, European Commission, DG XII.

Gilbert, N.G., P. Ahrweiler and A. Pyka (2001), *Understanding Innovation Networks thorugh Simulation*, Conference Contribution for EuroSim 2001.

Guy, K., J. Clark, K. Balazs, J. Stroyan and E. Arnold (1998), *Strategic Options for the Evaluation of the R&D Programmes of the European Union*, November 1998, prepared for STOA: http://www.technopolis.co.uk/reports/stoa.

Holland, J.H. (1992), *Adaptation in Natural and Artificial Systems*, A Bradford Book, Boston, MA: MIT Press.

Kowol, U. and W. Krohn (1995), *Innovationsnetzwerke. Ein Modell der Technikgenese*, in W. Rammert, G. Bechmann and J. Halfmann (eds), *Jahrbuch Technik und Gesellschaft 8*, Frankfurt/M. and New York: Campus, pp. 77–105.

Pyka, A. (1999), *Der kollektive Innovationsprozeß – Eine theoretische Analyse absorptiver Fähigkeiten und informeller Netzwerke*, Berlin: Duncker & Humblot.

Schumpeter, J.A. (1912), *Theorie der wirtschaftlichen Entwicklung*, Berlin: Duncker & Humblot.

Ziman, J. (ed.) (2000), *Technological Innovation as an Evolutionary Process*, Cambridge: Cambridge University Press.

PART FOUR

Conclusions

9. Conclusions

Günter Küppers and Andreas Pyka[1]

The overall lack of co-ordination and insufficient co-operation between knowledge producers and those utilising knowledge of technological relevance, that is, the laggard technology transfer, is one of the reasons for European deficiencies in innovation performance (European Community 1995). What exactly hindered previous innovation systems from performing their task? One of the main causes lies in the difficulties and failures of technology transfer in relatively closed systems where there are heterogeneous rationalities at work which create compatibility problems at their borderlines and hinder the respective technologies from passing over. For example, the financial investments of private companies, as well as the public funding of research, often lead to diverse rationalities. Patent rights, rules of commercial confidence and membership of international big firms guide the performance of private companies and, in doing so, restrict the efficiency of co-operation with governmental funding agencies. Companies do not invest primarily in the competitiveness of their (national or regional) location but in the (international) competitiveness of their enterprise. National interests regarding science and technology policies coming from government are only supported through strategic *ad hoc* alliances by different firms: they have to be compatible with the economic strategies of the companies. These alliances are easily abandoned, which often results in the withdrawal of industrial companies from technological encouragement programmes.

Another kind of borderline problem arises directly between universities and firms: on the one hand, scientific results cannot be easily implemented in existing structures of application contexts, which causes various integration problems, often resulting in the exclusion of new technologies ('knowledge without user'). On the other hand, concrete research demands arising out of application areas are refuted by scientists because of their lack of 'scientific relevance'. Neither universities nor semi-private research institutes have the necessary infrastructure to respond to the innovation pressure experienced by commercial users. Furthermore, academic researchers are not interested in performing the necessary non-scientific tasks (which need special qualifications) like maintenance, routines and services concerned with new technologies.

Quite similar borderline problems are also very likely to emerge between different populations of firms; that is different industries, but also even within a single industry. Technological spill-over effects – both intra-industrial as well as inter-industrial – are often severely restricted due to the 'non-invented-here syndrome' or missing absorptive capacities of the involved actors. Consequently, the huge potential for cross-fertilisation effects between seemingly different technologies are not explored or even exploited.

The last example of borderline problems to be given here focuses on the difficult relationship between the area of knowledge production and the public. Societal participation needs are strongly articulated by the planning and controlling efforts of governmental funding agencies as the politically legitimised representatives of public decision making. In recent decades, researchers have not only had to face a growing demand for legitimation, explanation and justification from funding agencies, users and the public; faced with these actors, it is more and more difficult to represent and carry out inner-scientific aspects and criteria of quality control. Instead, the various actors involved claim that scientists should undertake moral and juridical responsibility for their products after R&D. Public discussion and dissemination of research results is difficult: researchers in universities and in private companies, policy makers, representatives of commercial enterprises and other relevant actors use different languages, follow different functional logics and belong to institutionally separated contexts of action. What kind of knowledge is known or relevant, what becomes an 'issue', depends on chance or on self-organising processes within existing communication structures. Although – or because – knowledge seems to be difficult to disseminate, needs for participation, control and regulation in the field of public discussion arise, and lay their claim to taking part in the production of 'relevant' knowledge; existing modes of knowledge production cannot answer these needs.

These borderline problems within highly differentiated innovation systems have caused strong internal problem pressure which it has not been possible to neutralise by conventional strategies of further functional differentiation. Instead, they have provoked an evolutionary change in the area of knowledge and technology production. In overcoming the strict differentiations and borders of traditional innovation systems in order to deal successfully with arising innovation problems, diminishing competition deficiencies and accelerating innovation processes in research and technology development, a new means of knowledge production has structurally emerged; this can be characterised by the 'innovation network' metaphor.

Under the title of *Innovation Networks: Theory and Practice* the results of an interdisciplinary project, analying the structure and the dynamics of such innovation networks have been presented. Summarising, three major results have been obtained:

First, it could be shown that the concept of self-organisation is a fruitful tool to analyse the structure and dynamics of innovation networks. Because of turbulent markets and the complex dynamics of science and technology, the construction of innovations is subject to uncertainty: a lack of knowledge with respect to technical feasibility, economic success and social acceptance. The aim of an innovation network is to reduce this uncertainty through the co-operation of actors with different areas of competence in order to produce the missing knowledge. Hypotheses about this function are established and checked by experiments like technical tests of components or prototypes. Hypotheses about the internal organisation of the network are checked by social experiments with different forms of co-operation. The circular relationship between these hypotheses and the experiences resulting from their conversion yields to a value of its own where hypothesis and experience reproduce each other as a solution of technological as well as social problems of innovations.

Second, the different case studies analysed within this project have been chosen strategically with the aim of covering a wide variety of existing innovation networks. The subject of the case study on the Mobile VCE, is a 'virtual centre of excellence' in the area of mobile and personal communications (see Chapter 3 in this volume) It was set up in 1996, as a consortium involving seven UK universities and almost all the major European companies active in mobile communications. While Mobile VCE undoubtedly 'worked' according to the terms in which it was set up, this does not necessarily mean that VCEs are a useful policy tool. In comparison, another VCE for digital television technology, the Digital VCE, was very much less successful. We have found two sorts of alliance supported by Mobile VCE which are specific (though not necessarily unique) to VCEs as a funding mechanism and are of interest as a policy tool: the virtual links of the research network, and the self-perceived identity of the industrial network. The virtual research network in Mobile VCE encouraged inter-institutional work among research associates, as well as creating links between researchers and the industrial representatives involved in research management activities. The political aims of creating a virtual centre included the creation of a centre of excellence (by bringing together isolated research teams) and the encouragement of industrial-academic collaboration by providing industry with access to a wide range of the best academic work in relevant fields. Both these aims are better served by a VCE than by other policy initiatives aimed at forming research consortia. A VCE, it might be said, brings together two networks (the academic and the industrial), not merely individual actors.

Although Mobile VCE is largely judged a success, the VCE mechanism has not been reproduced beyond Mobile and Digital VCEs. One of the problems demonstrated by Digital VCE is the difficulty of identifying appropriate strategic sectors, where a critical mass of industrial interests can be enrolled in support of a virtual research centre. However, other industrial sectors might well be appropriate. The main problem for the future of the mechanism seems to lie in the broader selection environment, that is, the political priorities of the various funding agencies. The VCE mechanism may, for example, be held to offend against principles of competition – both in terms of encouraging competition between leading research teams, and in expecting the market (i.e. industrial funding) to take over from public funding. The latter has been a concern of the UK Engineering and Physical Sciences Research Council (EPSRC), for example, although it has supported Mobile VCE's second three-year research programme. On the other hand, European funding priorities emphasise support for industrial development as the output of collaborative research projects: Mobile VCE offends against this, in that it has relied on being at least marginally pre-competitive in order to win the combined support of a competitive industry.

In the so-called biotechnology-based innovation sectors a high frequency of co-operative agreements has been observed since the end of the 1970s. (see Chapter 4 in this volume). The main actors in these industries are Large Diversified Firms (LDFs) with large shares of already established markets and Dedicated Biotechnology Firms (DBFs), usually young start-up companies and university spin-offs with a strong knowledge base in the fields of biotechnology, as well as public research institutes and university laboratories. The asymmetric distribution of economic and technological capabilities and competencies is considered an important reason for early co-operative R&D efforts. Almost 20 years after biotechnology becoming a widely applied technology, already established Large Diversified Firms have developed considerable competencies and capabilities, and some of the dedicated biotechnology firms have become large companies with several thousand employees and a considerable economic power. Despite these developments there is still an ongoing trend in co-operation. Accordingly, the motives for engaging in networks have changed from the so-called complementary assets to the exploration of a broader range of the research horizon using innovation networks as an extended workbench in R&D.

For a better understanding of this development a simulation model of the evolution of innovation networks was set up. As it is an applied simulation exercise focusing on working out developments of a concrete sector, in the conceptualisation of the model much emphasis is placed on the characteristic features of this industry. Obviously the implementation of the model sketched here in the sense of a *history friendly model* was not an easy endeavour. The first step therefore was to analyse a prototypical case which makes it possible to detect the interactions of the numerous mechanisms and interrelationships.

In a second step the results of the simulations are compared to developments in the real world by applying concepts from graph theory which allow the analysis of overall network dynamics. Although there are still some significant differences between the artificial evolution of network structures and the real world networks, the results look promising, as they are able to reproduce at least qualitatively some developments which can also be observed in reality. The simulation facilitates the analysis of different scenarios, showing the influence of different environments as well as of policy measures aiming at the establishment of these new biotechnology-based industries.

In general, the model matches a number of the observed features of innovation in these sectors. Going through this analytical exercise has significantly sharpened our theoretical understanding of the key factors behind salient aspects of the development of networking in the biotechnology-based sectors and contributed to a more general understanding of innovation networks in other sectors.

The Knowledge-Intensive Business Services (KIBS) case study examines the construction and co-ordination of innovation networks supporting the development and diffusion of e-commerce and the particular role of knowledge-intensive services within these innovation networks. (see Chapter 5 in this volume) The empirical work in this study has focused on one particular type of KIBS, the professional web company and its interaction with its business clients. The development of an effective web-site is an essential prerequisite for companies wishing to engage in e-commerce. It is a misleading, but oft repeated, observation that basic web-pages are easy to design and that the syntax of HTML is relatively easy to learn. The construction of a website involves much more than the writing of a few web pages. Rather, the professional web companies contacted by this report offer their clients bespoke packages that entail the putting together of a whole series of hardware, software, programming and design elements in order to create a fully-functional e-commerce website. In other words, web companies are another type of integrator – a package integrator.

Another important role played by professional web companies is the education of clients, particularly those with little or no previous experience, about the potential and current limitations of e-commerce. Unfortunately, the information asymmetry that exists between providers and clients means the latter are potential prey for 'bad sellers'. These will undercut the competition in order to make short-term private gains while delivering sub-standard products to the end-user. By creating dissatisfied buyers who are unlikely to return to the market, these outfits are simultaneously 'poisoning the well for others'. The establishment of good selling practice is therefore essential if Gresham's Law – that bad 'selling practice' drives out the good – is to be avoided.

From the perspective of the internet, where customers and firms interact in e-commerce, a number of important issues arise when considering the role of political institutions within e-commerce. Particularly prominent are issues relating to the areas of access and social exclusion, data protection, trust relations, competition policy, standards-setting and IPR.

Large technical systems require a new sort of competition policy in which the spotlight is placed on inter-market relationships as well as intra-market conditions. As the DoJ vs. Microsoft court case in the USA has highlighted, monopolistic control of one area can confer significant leverage in another area, possibly stifling competition. Traditional perspectives on standards-setting also need to be revised. Innovation in network technologies is typically associated with several layers of standards that concern performance, design and interoperability. These cut across the traditional institutional demarcation of *de facto* market-driven standards and *de jure* standards set by government-sponsored bodies. The standards game is at the heart of the strategic battle being fought between the two rival networks of Microsoft and AOL-Sun-Oracle. Policy requires the development of a new, integrated perspective on standards-setting.

The case study on the role of inter-institutional networks for the innovation diffusion of Combined Heat and Power technology (CHP) and the transformation of energy supply systems is composed of three national studies of the developments in the UK, Germany and the Netherlands, covering mainly the period from the early 1980s to the mid/late 1990s (see Chapter 2 in this volume). This case study showed that setting up innovation networks alone is often not enough to establish an innovation. There is a high risk that the innovative effort will stop at an experimental stage and never reach wider application. In many cases, this is due to a lack of 'embedding' of the innovations in a compatible structural context. It may require structural and contextual changes to enable the wider uptake of an otherwise promising innovation.

Regulatory and policy frameworks are critical elements of this context which can be influenced by policy. In fact, measures aiming to establish conducive framework conditions for innovations in a particular field like energy supply can be at least as efficient for creating innovations as the bottom-up operation of innovation networks. In the CHP case study, structural and regulatory barriers were more constraining than technical and economic ones. The regulatory reforms in energy supply therefore opened up new opportunities for innovative energy supply solutions to emerge, though not necessarily to succeed. Most notably the UK case also showed that a strategy that relies solely on transforming the framework does not necessarily lead to the kinds of innovative outputs desired from a policy perspective.

The most effective way of managing innovation diffusion processes in complex problem areas (such as energy supply) seems to consist of a combination of networking initiatives that aim at stimulating self-organising forces and creating a variety of novel solutions, and of adjustments of the framing regulatory and policy context to define a corridor of desired results.

The CHP cases also showed how important organisational innovations are in addition to the technological changes. The existence of potential carrier organisations that have an intrinsic interest in the kinds of innovation that are expected to emerge turned out to be decisive for establishing innovation networks. Their role is to enable and facilitate the dissemination of and learning process about good practices with an emerging technology and thus the establishment of an information network. They can be set up around established and respected key actors (for example, industrial associations), but in view of their supporting function for the policy process it may be even better to set up an independent body.

The feasibility of both organisational changes and adjustments of the regulatory and policy context require networks that address these political aspects of innovation diffusion in large socio-technical systems to be in place. It is thus important to consider the wider political barriers to innovations as well and take them into account in the design of innovation policy.

Third, in the operationalisation of a newly developed theory of innovation networks the chapter on the simulation model introduces a multi-agent system that makes it possible to test political parameters and derive practically oriented conclusions from simulation experiments based upon them. Innovation is dynamically initiated as the successful implementation of new ideas. In the process, the agents modelled – such as companies, policy makers, research laboratories, etc. – use their specific knowledge to produce artefacts that are presented as potential innovations. The success of these artefacts is determined by an Innovation Oracle that evaluates the artefacts in accordance with criteria not available or unknown to the agents. In order to enhance both their strategies and their results, the agents are able to improve their own knowledge and abilities by means of incremental or radical changes to their knowledge base, or to opt instead for co-operation and partnership, drawing on external sources of knowledge.

The analysis of the simulation model, which is designed as a platform model allowing the implementation of various case studies, shows how the various parameters and their respective combinations produce model behaviour that reveals qualitative similarities to empirically located innovation networks outlined in the chapters on the case studies. The model experiments furthermore allow for generalisations with respect to individual aspects of an innovation policy aimed at promoting innovation networks.

Because innovation networks integrate structural and dynamical perspectives, a radical change in evaluation theories, methods and tools is required (see Chapter 8 in this volume). It is not only the product and its impact which is the focus of evaluation, it is also the production within a network which requires new concepts of evaluation. Problems of the performance of the network, the integration of the context of application, the problem of social accountability pose new questions for evaluation. The perspective developed within this project calls for a new evaluation approach for RTD programmes which encourage innovation networks, such as the Framework Programmes of the European Commission. Taking those as a model, the approach refers to goal attainment questions arising within the European Framework Programmes and is mainly directed to EU policy makers, though it can be used for national and regional RTD evaluation as well.

One obvious difference to 'traditional' Mode 1 evaluations is the variety of relevant agencies which need to be considered within innovation networks – business organisations, consumers, academics and political institutions. This is consistent with Georghiou's call for an extension of the scope of policy evaluations. The simulation model provides a clearer picture of how the constituent elements of these innovation networks interact if policy makers are to have a proper appreciation of the impact of EU-funded projects and programmes. In addition to agency selection, other key aspects requiring investigation include the choice of dimension and question selection; that is, which issues are to be investigated and what type of questions are to be asked. Each clearly affects the type of research conducted and its focus, and hence, the type of results that are generated.

The aim of this book is to bring together the sociological, the economic as well as the political literature on innovation networks in order to develop a general theory of the emergence and development of this, meanwhile ubiquitous, phenomenon of innovation networks in knowledge-based industries. Besides this interdisciplinary feature, the theory of self-organisation is invoked in a transdisciplinary way, as a theoretical concept which has shown its fruitfulness in the analysis of complex and path-dependent processes. The model of innovation networks we developed is not immediately oriented on a phenomenological approach but on the understanding of the internal structures and dynamics of innovation networks. Accordingly, the general model has to be considered as a platform which should allow the description as well as the analysis of a large number of classes of different empirically relevant innovation networks. The identification of the rich space of possible outcomes of the model allows the distillation of processes crucial for almost all forms of innovation network as well as of those which are very specific to singular empirical cases. This specific methodology on the one hand should allow a better theoretical understanding of the factors, processes and dynamics behind the observed phenomena. On the other hand, the model can also be understood as a policy tool, allowing experimentally the evaluation of different policy designs.

NOTES

1. We acknowledge the contributions of the authors in this volume.

REFERENCES

European Community (1995), *Green Paper on Innovation* (COM (95) 688), Brussels: EC.

Author Index

Subject Index